World City Network

Contemporary globalization describes a situation in which important social relations are increasingly stretching worldwide, transcending national boundaries. But how are these transnational flows and connections organized if not through countries? The most common answer to this question is that cities are the organizational 'hubs' of globalization.

World City Network interprets cities as global service centres. With the advent of multinational corporations, the traditional urban service function has 'gone global'. In order to provide services to globalizing corporate clients, the offices of major financial and business service firms across the world have created a new network. It is the myriad of flows between office towers in different metropolitan centres that form the world city network. Through an analysis of the location strategies of 100 leading global service firms across 315 cities, this book assesses cities in terms of their overall network connectivity, their connectivity by service sector, and their connectivity by world region.

Peter Taylor's book provides a first comprehensive, systematic description and analysis of the world city network as the 'skeleton' upon which contemporary globalization has been built. His analyses challenge the traditional view of the world as a 'mosaic map' of political boundaries.

Peter J. Taylor is Professor of Geography at the University of Loughborough, UK, and Research Professor at the Metropolitan Institute, Virginia Tech, USA. He is an Academician of the Academy of the Social Sciences (UK) and the Association of American Geographers' distinguished scholar for 2003.

World City Network

A global urban analysis

Peter J. Taylor

Routledge
Taylor & Francis Group

LONDON AND NEW YORK

First published in 2004 by Routledge
11 New Fetter Lane, London EC4P 4EE

Simultaneously published in the USA and Canada
by Routledge
29 West 35th Street, New York, NY 10001

Routledge is an imprint of the Taylor & Francis Group

Typeset in Times by Florence Production Ltd, Stoodleigh, Devon
Printed and bound in Great Britain by TJ International Ltd, Padstow, Cornwall

British Library Cataloguing in Publication Data
A catalogue record for this book is available from the British Library

Library of Congress Cataloging-in-Publication Data
Taylor, Peter J.
  World city network: a global urban analysis/Peter J. Taylor
    p. cm.
  1. Cities and towns. 2. Intergovenmental cooperation. 3. Globalization.
  4. Information networks. 5. Service industries. 6. Economic geography.
  I. Title: Global urban analysis. II. Title.

  HT119.T39 2003
  307.76–dc21   2003005320

ISBN 0–415–30248–X (hbk)
ISBN 0–415–30249–8 (pbk)

For Tom

Contents

 **Preface**

This book has been several years in the making and has had, for a single-author text, an unusual degree of input from others. This happenstance originates from my creation of GaWC (Globalization and World Cities Study Group and Network) at Loughborough University in early 1998. Set up to counter the dearth of data on inter-city relations, this virtual centre (www.lboro.ac.uk/gawc) has grown immensely to constitute a vibrant research group at Loughborough interacting in a network of world city researchers across the globe. Appropriately, it is an institute of globalization to study globalization. In practice this means that I have been able to do my research in conjunction with many other researchers. Hence in my list of personal acknowledgements, all those named are co-researchers who have contributed directly to my work and therefore to this book: Jon Beaverstock (Loughborough), Richard Bostock (City of London), Ed Brown (Loughborough), Gilda Catalona (Calabria), Ben Derudder (Ghent), Troy Gravitt (Virginia Tech), Michael Hoyler (Heidelberg), Phil Hubbard (Loughborough), Stefan Kratke (Frankfurt an der Oder), Kathy Pain (Loughborough), John Short (Maryland), Richard Smith (Leicester) and David Walker (Loughborough). In addition, many other contributors to the GaWC web site have had an indirect influence, and to all GaWCers (pronounced with a hard 'c') I record my thanks.

The research reported in this book is only one strand of GaWC research, the quantitative work. This is extensive research into the patterning of the world city network, and it provides the context for intensive (qualitative) studies of inter-city processes. Much of the work has involved large-scale data collection and this was possible only through the support of the Economic and Social Research Council covering several projects. Without the research council monies this book would never have been written. Actually, there are persons who think that it should not have been written. In writing up this research as articles, I have experienced, for the first time in my career, rejections by geography journals. Referees who confuse empirical with empiricism and quantitative with positivism have referred to the work on several occasions as 'old-fashioned'. Clearly, these are immature people who have never worn flares. All the papers so aggressively dismissed by British geography journals have been published elsewhere, with few or no changes, in urban studies and US journals. The lesson is obviously not to submit papers to British geography journals; the irony is that the cultural turn emphasizes difference but seems not to respect it in practice.

Half this book was written while I was at the Metropolitan Institute, Virginia Tech, and it was completed on returning to the Department of Geography, Loughborough University. Thanks are due to Rob Lang and Jennifer LeFurgy, and Morag Bell, for promoting two such pleasant and productive working environments as northern Virginia and Loughborough respectively. I must say a special thanks to Mark Szegner and Clive Cartwright, who produced the many diagrams, and to Jacki Bowyer, who oversaw the final production.

The book is dedicated to my new grandson, Tom. My daughter assures me that he is not named after my first boyhood hero, Tommy Lawton (Notts County and England!), but I choose to think otherwise.

Peter Taylor
Tynemouth, December 2002

Acknowledgements

Figure 1.5 is reproduced from *Development and Change* (17(1)) with permission of Blackwell Publishing (Oxford).

Figure 2.1 is reproduced from T. R. Lakshmanan and P. Nijkamp (eds) *Structure and Change in the Space Economy* with permission of Springer-Verlag GmbH & Co. KG (Heidelberg) and Roberto Camagni (Politecnico di Milano).

Prologue: the second nature of cities

In 1945 Chauncy Harris and Edward Ullman published their classic study 'The nature of cities'. Over the next few decades this article achieved the status of a standard in its field, copiously cited by researchers and reproduced, in whole or in part, in numerous textbooks and readers for teachers. Agnew (1997: 5) records its presence in seven anthologies of urban studies covering the period 1951–70. I focus on two features of this classic article here: its overt multidisciplinarity and the selectivity in its use.

In a period when social science disciplines were establishing themselves as separate knowledges, the Harris and Ullman (1945) article stands out as a multidisciplinary project. Harris and Ullman were geographers who chose to publish their article in a leading sociological journal. The article is widely remembered for its famous composite diagram showing concentric, sector and multiple nuclei models of cities. Of these three basic models of the internal structure of cities, one was from sociology, another was from land economics, and they added their own geographical proposal. And this multi-disciplinarity is very proper: cities are inherently multifaceted, therefore the idea that any one 'specialist' social science can encompass this complexity is foolhardy. This volume comes out of the same disciplinary stable as Harris and Ullman and navigates through a similar contemporary multidisciplinary subject matter. My chief guiding lights have been an economist (Jane Jacobs), a planner (John Friedmann) and two sociologists (Saskia Sassen and Manuel Castells), aided and abetted by many fellow geographers. In these more 'global times', I have personally found a curious geographical division in the multidisciplinarity: presentations of the ideas upon which this book is based in continental Europe have been largely to geography audiences, the few papers delivered in Britain have been largely to planners, and nearly all presentations in the United States have been to sociologists. No matter, wherever and whatever the urban researchers, all are cognizant of the need to work across disciplines and not within them. This book is written in precisely this spirit.

In a more negative vein it has to be noted that the use of the Harris and Ullman article has been very selective. The renowned three-models diagram is taken from the second section of the paper, entitled 'Internal structure of cities'. The first section, entitled 'The support of cities', is rarely referred to in the literature; until I read the original article I had always thought that it dealt just with the internal relations of cities. Somehow this classic work has been celebrated for just half of its argument. This is consistent with much of the urban literature wherein an 'urban theory' is developed that is invariably about the internal goings-on of cities. This neglect of the external relations of cities, which has varied markedly over time, is the subject matter of this book; I call it the 'second nature of cities'. After all, it is 'second nature' to cities to be connected to one another.

The external relations of cities are not an optional 'add-on' for theorizing the nature of cities. Connections are the very *raison d'être* of cities. There is no such thing as a single city operating on its own; cities come in packs and much research has gone into finding out how groups of cities are organized. Of course, this research has also been multidisciplinary: in sociology C. J. Galpin, in geography Walter Christaller and in economics August Losch all provided descriptions and models of cities and towns as service centres organized into urban hierarchies. Although just as influential as the internal structure models, these external relations models have not always been part of how social scientists viewed the nature of cities. While they paid lip-service to ideas such as cities being the 'crossroads of society', relations between cities came to play second fiddle to relations within cities in urban studies. This is partly because the external models were appropriated to create a space–economy framework of locational models of levels of economic activity: Christaller *et al.* explained tertiary-level activities (as location of services models) to complement other models dealing with primary activities (agricultural location models) and secondary activities (industrial location models). Thus under the leadership of regional economist Walter Isard, this 'locational synthesis', by focusing on the services and not the cities, was partly responsible for removing the formal study of city external relations from 'urban theory'. This book is written exactly in opposition to this outcome: it is a reassertion of external relations as the second nature of cities.

As with all generalizations, there are important exceptions, and in this case they are within geography. While Isard (1956) was creating his economic synthesis, Brian Berry (1964) was arguing for a city-orientated geographical synthesis as 'cities as systems within systems of cities'. In this famous formulation the internal and external relations are fused into a single framework for describing the nature of cities. And this is, of course, how it should be. With obvious implications for urban, regional and national planning, Berry's thinking remained important in policy circles but was somewhat eclipsed in urban geography itself: by the 1980s one comprehensive review of current research was able to focus solely on the internal relations of cities (Bassett and Short 1989). Globalization, and the obviously important role of cities in this process, has changed all this. Originally in reference to 'the new international division of labour', cities as 'command centres' in a globalizing world-economy were initially modelled by John Friedmann from the 'urban systems' school in urban planning. Following his earlier work on national urban hierarchies, Friedmann (1986) proposed a world city hierarchy, and it is from this point that a world cities literature begins to develop in a big way. This book fits into this literature but claims an unlikely niche.

Returning to ideas of hierarchy means bringing back concern for the external relations of cities. Obviously, cities are 'world' cities because they have external relations that are transnational or even global in scope. Although our image of these cities is a very static representation – the ubiquitous tower office blocks – this built environment houses numerous nodes that send out and receive trillions of messages (how many services are carried out) through cyberspace every day. World cities are the crossroads of this globalization. But these external relations have proved to be very difficult to research. Hence although the world city literature is premised upon giving due weight to the external relations of cities, in practice most studies have either focused upon a single leading city or made comparisons between two or three leading cities. But comparative analysis is not a relational analysis. For the latter we need to have information on the connections, links, flows – in short, the relations – between cities. And here is the rub: such information is extremely hard to come by. Thus it is 'unrelational' research on cities that continues to dominate urban studies even within the world city literature. The second nature of cities continues to be neglected. It is the purpose of this book to begin the task of confronting this neglect and producing a relational study of world cities.

Moving from the internal nature of cities to their external nature brings to the fore many new theoretical issues (see Amin and Thrift 2002). In a particularly interesting 'theoretical taxonomy' of research on 'global cities', Camagni (2001: 103) identifies four types of study: 'City as Cluster', City as Milieu', 'City as Symbol' and 'City as Interconnection'. This book comes into the last of these categories, and within this type the focus is largely on economic connections. Thus the research is very big geographically – global – but very narrow in topic. It is a detailed study of a large number of contemporary financial and business firms with a global reach. This is where the empirics lie, but interpretations take the argument further into historical and political realms.

The text is divided into four parts: Relations, Connections, Configurations and Suppositions. In Part I the relevant literature on cities and globalization is reviewed, critiqued and borrowed in such as way as to provide a basis for the rest of the study. In particular, the urban economics of Jane Jacobs is found to be the best starting point and I link this to a broader world-systems framework. Parts II and III are the empirical core of the book wherein I present what the subtitle calls a 'global urban analysis'. This is based upon a specific concentualization of the world city network that guides data collection to produce a large data matrix that forms the input to most of the subsequent analyses. These involve both ego-centric analyses in Part II – looking at the connectivities of cities one by one – and network-level analyses in Part III – delineating global configurations. Finally, in Part IV a preliminary theoretical excursion into the geohistorical meanings of the global urban analyses is presented. Locating the analyses as a present between pasts and futures, policy-practical and theory-ontological issues are discussed.

The purpose of this prologue is to provide a flavour of the book and to ensure that readers understand where I am coming from. This is important for what is essentially an empirical monograph. In addition, as just indicated, the final part of the book says something about where I hope to be going with this work. It is worth stating this here in general terms so that readers may bear this in mind as they work their way through the analyses. 'Urban problems' dominated urban research in the second half of the twentieth century to create what might almost be termed a 'hate literature' on cities. In fact, as has been pointed out many times, these so-called urban problems were actually general problems of modern capitalist society expressing themselves in urban locales. However, this conceptualization of cities as 'problems in search of solutions' did contribute substantially to the 'internalist turn' in urban studies reported above. The message of this book is that we need to begin to learn how to love our cities again. The current 'love of country', with all that that implies militarily, is far too dangerous for a globalizing world. Countries, aka states, have an appalling record for destroying cities. World development has been largely possible in spite of the actions of competitive states. Citizenship, aka nationality, needs to revert to co-operative cities under conditions of contemporary globalization. A choice between a corporate dystopia and a citizens' utopia is much better than choosing between little and large Armageddons. This is why the second nature of cities is so important.

 **Relations**

1 Inter-city relations

Dependence on a network rather than on the servicing of an environing region, or a wider hinterland, existed for a few exceptional cities in the past, but now it has become the general rule for the majority of substantial cities anywhere.
(Gottmann 1989: 62)

This chapter introduces relational thinking about cities. The relational thinking I am interested in is not that which focuses upon one city and asks about its internal relations or how it relates to its hinterland. In this text, cities will always be referred to in the plural because the focus is on inter-city relations, on dependencies and interdependencies between cities. These can never be adequately understood by starting from just a single city, because inter-city relations form configurations of connections across many cities. In addition, the configurations I concentrate on are those that directly inform a material understanding of cities, how cities work together as economic entities. And although the book as a whole aspires to contribute to debates concerning the nature of *contemporary* globalization, issues that arise are by no means unique to these 'global times': ultimately, what I present is a geohistorical interpretation of world cities in networks. In this chapter I provide some necessary background – historical and literature reviews – for considering the different forms that inter-city relations might take.

The chapter is divided into three substantive sections arranged in chronological order. As this is a geohistorical study, it is vital that relations between cities are considered before the advent of the nation-state. One of the theses developed later in the book is the idea that because current social knowledge is so state-centric in nature, other non-territorial arrangements tend to be missed. Here I use a little historical knowledge on inter-city relations to illustrate alternative patterns of spatial organization. This is the task of the first section, which considers three well-known examples of inter-city relations that precede the modern 'nationalization' of cities. The growth of urban studies from the mid-twentieth century was almost entirely state-centric and the second section looks at how the research on inter-city relations as 'national urban systems' was conducted. One concept of inter-city relations totally dominated this school of research: 'urban hierarchy'. Since this concept became endemic to relational thinking on cities, it has influenced how world cities have come to be conceptualized, not least because researchers from the former national-scale literature have made key initial contributions to the literature on how cities are organized under conditions of contemporary globalization. World cities, or global cities, are the topic of the third section, where I highlight the seminal contributions of Friedmann, Sassen and Castells. Since this is also the literature to which this book contributes, in a short concluding section Castells' idea that there is a new spatial logic for understanding inter-city relations is illustrated. The remainder of the book is set firmly within the logics of his spaces of flows.

Cities as process

In his treatment of cities, Manuel Castells (1996: 386) considers them to be 'not a place but a process'. By this he means 'a process by which centres . . . are connected in a global network'. This is a very important idea and it should not be restricted to conceptualizing today's global cities. If cities in general are a process in this Castellian sense, then it follows that there can be no such entity as 'the city', meaning a single city. Cities are networks; individual cities within networks have hinterlands but the latter are never sufficient to create and maintain an 'isolated city'. The latter is a myth that appears in archaeology in the search for the first city, sometimes called the 'mother city', in a region. There is no such process of one city initially standing alone but whose success then diffuses the idea of 'the city' throughout a region. It is city networks that are constructed in any urbanizing region: there are always multiple 'mother cities'. Cities need each other. Similarly, the single city at the centre of von Thunen's famous 'isolated state' is an abstraction defining a situation that could not happen. Even capital cities need other cities.

Castells' portrayal of cities as a networking process appears to be innovative in contemporary urban studies only because researchers have become accustomed to seeing cities as urban places, complete with boundaries, within their respective countries. Somehow the idea of the connectiveness of cities has been largely lost and 'the city' is alive and well in modern social science. Reading the historical literature, I get the very strong feeling that understanding of cities as process was far more prevalent – indeed, may often have been the norm – before the advent of the nation-state. After all, as Jean Gottmann (1984) has asked, how else would cities be organized except as networks? I use three key historical examples of city networks to ease the reader into thinking about cities as networks.

Abu-Lughod's transcontinental archipelago of cities

The starting point for this geohistorical excursion towards contemporary world cities is 1250–1350. It was at this time that the trading systems of Eurasia became integrated to an unprecedented degree: from China to Europe, commodities were exchanged in a transcontinental 'world system' that preceded Wallerstein's (1979) modern world-system by two centuries. Whereas the latter begins the era of European world dominance, in the former 'system' there is no dominant region and Europe is arguably the least developed economically. This intriguing circumstance – world integration before European expansion – has attracted the attention of Janet Abu-Lughod (1989) in a project to understand alternatives to the extreme geographical concentration of power exhibited by the contemporary world-system.

To describe the thirteenth- and fourteenth-century 'world system' Abu-Lughod (1989: 14) chooses 'to focus on cities rather than countries' in order 'to trace the connections' between the most economically intense parts of the system. The result is the depiction of a vast archipelago of cities constituted as eight overlapping regional city networks linked together to form a transcontinental pattern of connections. The basic configuration of this archipelago consists of two east–west routes, an overland caravan passage in the north and a sea passage in the south (Figure 1.1). Tracing these routes, I will start in the far north-west of the archipelago and begin with Bruges, the centre of what Braudel (1984) calls the 'northern pole' of Europe, linked to the Italian 'southern pole' represented by Genoa and Venice. From the Mediterranean there are three routes east: through Constantinople to the overland route, through Alexandria and Cairo to the sea

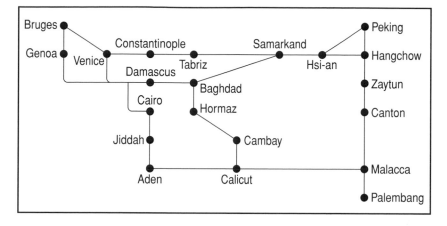

Figure 1.1 A transcontinental archipelago of cities *c*. 1300 (derived largely from Abu-Lughod 1989)

route, and through Acre and Damascus to Baghdad with its alternative links with both land and sea routes. The overland route goes through central Asian cities (Tabriz and Samarkand) to Peking, the sea route through Arab and Indian cities (Aden, Calicut) and on to Malacca (for the Spice Islands) and Hangchow in China, the greatest city of the era.

What was the nature of this vast configuration of cities? What were the processes that created it and held it together? Although Abu-Lughod (1989: 13) refers to 'a common commercial network of production and exchange', it does not follow that Figure 1.1 describes a coherent network of cities – which is why Abu-Lughod calls the configuration an archipelago rather than a network. This is reasonable, because the distances, with travel times between cities measured in months, militated against integrated city network formation. There was no direct trade between cities from different ends of the archipelago; rather, middle cities operated as exchange centres, as great entrepôts in this world of merchants. Thus Arab, Persian and Indian merchants acted as go-betweens in the commodity flows as goods passed between different circuits of merchants on the journey between Europe and China. The end result was perhaps not a network of cities, but the overlapping regional city networks that Abu-Lughod describes do provide a glimpse of a worldwide configuration of inter-city relations that presages the contemporary world city network.

To understand the formation of city networks, therefore, it is necessary to delve into the regions that constituted Abu-Lughod's worldwide pattern. Fortunately, there is a comprehensive description of the European city network by Peter Spufford (2002) that I can use to show how these 'pre-modern' cities operated as a network. There was, of course, trading between towns and cities within Europe before its commercial revolution of the thirteenth century, but the volume remained low. Traders travelled between markets and fairs with their goods and money to buy cheap and sell dear in the traditional commercial manner. As trade increased, there was a critical point where quantitative change gave way to a qualitative change, and it is this that marks the commercial revolution: beyond adding extra mules to increase carrying capacity, new modes of trading were developed (ibid.: 19). This development involved creation of a division of labour within the trading enterprise. Instead of the single peripatetic trader there emerged three parties to transactions: sedentary merchants located in the leading cities; specialist carriers transporting goods between cities; and agents representing the sedentary merchants in cites beyond the home city. It is these new arrangements that converted myriad itinerant merchant paths and circuits into a network of cities.

The new trading activities were initiated in central and northern Italy through trading companies created to operate and finance the increasing trade. These were 'multi-branched companies' (Spufford 2002: 24) whose spatial organizations connected cities. Initially (in the twelfth century), agents were employed beyond Europe in Acre, Alexandria and Constantinople to organize the trade from the east, but by the end of the thirteenth century agents were widely distributed in cities across Europe as far afield as the Low Countries and Iberia (ibid.: 19, 24). With trading no longer a matter of personal bargaining between buyer and seller, new financial and business services evolved to facilitate the more complex trading practices. Indirect trading required book-keeping, an activity not necessary with personal dealing. A new 'commercial arithmetic' developed (ibid.: 30) to keep accounts and carry out audits. Also, with the trader no longer in direct possession of, and therefore protecting, goods, the need for insurance arose to spread the risk: maritime insurance dates from 1350 (ibid.: 31). And, of course, with greatly increased trade, transactions were no longer conducted through payment in gold or silver: bills of exchange both discontinued the need for bullion to accompany goods for trading, and greatly increased the amount of money in circulation. And behind all this servicing of trade there was a massive courier service between cities for moving bills of exchange, delivering instructions to agents, and sending information back to firms. The extent of the courier service by the late fourteenth century is shown in Figure 1.2. Notice the continuing concentration of the service in central and northern Italy. The quantity of the flows was quite remarkable: one firm in the late fourteenth century is recorded as sending 320,000 dispatches, with, for instance, 17,000 between Florence and Genoa and even 348 between the banking centres of Naples and London at opposite reaches of the network.

The importance of this example for later considerations is in understanding how cities are linked together in two ways: first, through multi-locational firms; and second, through

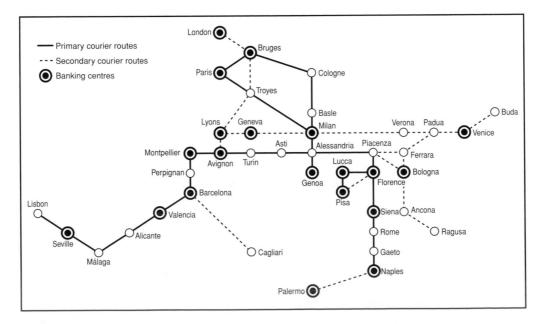

Figure 1.2 Inter-city courier services in late medieval Europe (derived from Spufford 2002)

a courier service that provided a network infrastructure. These types of connections will appear again in several contexts below.

The German Hansa 'community' of trading cities

Figure 1.2 provides the detail in the European sector of Abu-Lughod's archipelago (Figure 1.1), showing not just that the links between Venice/Genoa and Bruges took two routes, via Paris and via the Rhine, but also how the rest of Europe was integrated into the worldwide pattern of trade. Thus within the Europe of the times there was most certainly a network of cities but, curiously, this was most strongly institutionalized in the 'northern pole' of cities, represented so far by only Bruges. There was, of course, much more to trading in Northern Europe than just one city. Here is to be found a polit-ically organized city network that developed before the rise of the modern world-system. In fact, the 'Hanseatic League' is probably the best-known historical organization of cities as a network. I will use it to explore how city networks operate to find further hints for understanding city network formation.

Although 'Hanseatic League' is the common appellation for this network of cities, the main authority on the subject insists that it was never a 'league' of cities and prefers the concept of a 'community' of cities (Dollinger 1970: xx). This is because a league implies a precise organization, with a specific purpose that may be quite short term. In contrast the German Hanse was a nebulous organization, with fluid membership never firmly set-tled, but which existed for nearly five hundred years from the twelfth to the seventeenth century. Originally an association of merchants trading between eastern and north-west Europe, in the second half of the fourteenth century it evolved into a community of cities with between seventy and eighty cities constituting the key membership and another hundred or so as additional members (ibid.: 86–8). The general purpose of the Hanse was to promote the trade of its members by obtaining 'privileges' from trading partners in order to create economic monopolies for its members. Although it did sometimes resort to military intervention to get its way, the chief tool of the Hanse was trade blockade. Thus was Northern Europe 'trapped by a chain of supervision and dependence', creating a Hanse 'solidarity' based upon a strong 'community of interest' (Braudel 1984: 103–4).

The Hanse network comprised three parts. First, there were the Hanse cities them-selves, the home towns of the merchants. Second, there were the communities of Hanse merchants in cities where they obtained rights and privileges to trade. Third, there were other cities in which Hanse merchants traded but without any special privileges. The basic configuration of this organized network of cities is shown in Figure 1.3. The basic trade consisted of exchanging East European primary products (wood, wax, fur, rye and wheat) for North-West European cloth, wine and salt. To conduct this trade there were a string of Hanse ports along the Baltic and North Sea coasts at the mouths of rivers that linked to hinterlands of inland Hanse cities. At the centre of this network stood Lübeck, controlling the waterways (canal and rivers) linking the two North European seas. Lübeck was indisputably the central city of the Hanse – the *Hansetag* (diet) normally met there (Dollinger 1970: 92) – but with such long distances involved, the Hanse was also organized regionally, first into 'thirds' and then re-formed into 'quar-ters' each with their (sometimes changing) leading city. Each region defined a coast–hinterland sector: in the central region Lübeck was indisputably section leader; in the east, Visby and Danzig vied for similar status. The densest part of the network was in north-west Germany with two sections, one centred on Brunswick and the other first on Cologne and then on Dortmund.

Among the scores of settlements of German merchants in cities beyond the Hanse there were four key sites, *Kontore* in cities where they enjoyed special privileges,

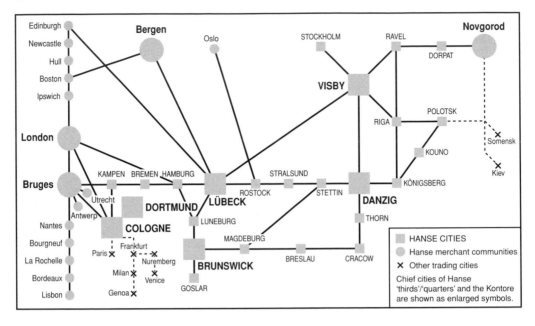

Figure 1.3 The German Hanse city network

including effective self-government (Dollinger 1970: 98). The most important *Kontor* was in Bruges, the leading city in the 'northern pole' of medieval European economic revival and trade expansion. The merchants of this *Kontor* dominated Hanse trade with the Low Countries but also claimed authority over smaller groups of German merchants in Atlantic ports as far away as Lisbon. Across the North Sea there was a second *Kontor* in London that served as the main base for English trade. In the far north, Bergen served as the home of the *Kontor* of merchants supplying Atlantic fish to Europe. In Eastern Europe the traditional trading centre of Novgorod was the site of the fourth *Kontor* of merchants but this was latterly challenged by separate associations of German merchants in other East European cities. The geographical outcome of all this trading activity was to create a cohesive city network from Novgorod through Lübeck to Bruges and London with extensions north to Scandinavia and south down the Atlantic coast. In addition, the German Hanse was connected to Mediterranean city networks from Cologne through the non-Hanse German cities of Frankfurt and Nuremberg to Milan and Genoa, and Venice respectively.

The process of creating this city network is quite plain. Remember that the German Hanse began as informal associations of merchants before it was consolidated into a community of cities. Thus formation of the city network was clearly the work of the merchants. The standard way in which city linkages were created and maintained was by merchant partnerships (Dollinger 1970: 166). Sharing the risk in trading enterprises involved bringing together leading merchants from several cities for a particular operation covering just a few years. Thus

> every great Hanseatic merchant was an associate in many partnerships which were linked together only through him. ... However extensive his business may have been, he figured less as the head of a great commercial undertaking than as a participant in a number of separate businesses.
>
> (ibid.: 167)

These are the agents of the city network formation; they link cities together in a myriad of partnerships that create the network. Braudel (1984: 103–4), as usual, sums up the situation beautifully: 'the same families – the Angermundes, Veckinghusens, Von Soests, Gieses and Von Suchtens – might be found anywhere between Revel, Gdansk (Danzig), Lubeck and Bruges. . . . All these links made for coherence, solidarity, habits in common and a shared pride.' Such a process defines an interlocking network wherein the nodes (cities) are connected through the activities of trans-nodal agents (the merchants). This way of understanding city network formation is how I investigate the process of contemporary world city network formation in later chapters.

One corollary of network formation through myriad links by agents is that the outcome, first, is very complex, and second, will inevitably encompass numerous contrary tendencies. Thus the depiction of the Hanse network in Figure 1.3 is a simplification in two senses. First, and most obviously, only a few of the most important connections are shown; to draw them all would make the diagram indecipherable. This is fine up to a point, but focusing merely upon leading connections misses the conflicts and contradictions in a network formation. This can be illustrated through English connections to the Hanse (Lloyd 1991). The connections shown in Figure 1.3 link all German communities in English cities to London. This shows a clear hierarchical tendency in this section of the network: the London *Kontor* brought all German merchant communities in England under its control, leading to, for instance, flows of delegates and merchants' dues to London (Dollinger 1970: 106). This made sense for Hanse negotiations with the English government. But the merchants outside London had their own trading patterns, allowing them opportunities to resist London interference. The example of Boston and its link to the Bergen *Kontor* is illustrated in Figure 1.3. These German merchants in Bergen and Boston were largely from Lübeck, whereas the London *Kontor* was dominated by Cologne merchants. In contrast, Great Yarmouth's German merchants were largely from Hamburg, Lynn had particularly close relations with Bremen merchants, and Hull merchants with Stralsund (Lloyd 1991: 42, 85–91). In addition, the German merchants in English ports traded to and from very many European ports across the whole network. Thus there was plenty of practical autonomy from the hierarchical tendencies focused upon London. The lesson of all this is that city network formation by innumerable interlocking agents was, and will most likely always be, perforce a multi-faceted process and outcome.

Braudel's world-cities

Both the transcontinental configuration of cities and its medieval European periphery reached their zenith about 1300, after which signs of economic decline appear. This is symbolized by the Black Death epidemic of the 1340s, which spread through the archipelago of cities from east to west. The severe demographic falls had profound economic effects, but this was not the only factor. Another factor was the disruption of trade from Europe to the east. More generally, this can be interpreted as the unfolding of Braudel's *longue durée*, a cyclical sequence of economic decline following two centuries of growth before 1300.

However the change is interpreted, the effects were real enough for the cities. The linking of cities described by Abu-Lughod (1989) breaks down as a single transcontinental archipelago after 1350. And in the European subregion, Braudel (1984: 117) refers to the economy as being 'reduced more than ever to "archipelagos" of cities'. The latter refers to a reduction in network connectivities that affected both the northern pole centred on Bruges and the German Hanse, and the southern pole centred on Genoa and Venice. However, in this restructuring it is the latter pole that fares best in the crisis, and it is here that the subsequent revival of the European economy is centred, a revival that ultimately culminates in new European-dominated transcontinental links which I call the modern world-system. Braudel (1984) charts a route from late medieval restructuring to the modern world-system using the concept of *world-city*.

Braudel (1984: 22) defines a world-economy as 'an economically autonomous section of the planet able to provide for most of its own needs, a section to which its internal links and exchanges give a certain organic unity'. There are 'ground rules' (ibid.: 25) through which activities of world-economies operate. One critical rule that Braudel identifies is that 'a dominant capitalist city always lies at the centre' (ibid.: 27). This city is the world-city for that particular place and time. These are cities with 'international destinies' (ibid.: 26), but more specifically they are at the 'logistic heart' (ibid.: 27) of the world-economy. There are two features of these world-cities that are important. First, they are not isolated, they need other cities; the latter may be subservient but they are necessary nodes in conduits of flows (ibid.: 27). Second, world-cities change as part of the development of a world-economy. The six hundred-year sequence Braudel describes for restructured Europe and the consequent modern world-system is Venice, Antwerp, Genoa, Amsterdam, London and New York.

Venice is the European city that suffers least in the medieval economic downturn and therefore becomes the world-city of a Mediterranean-orientated European world-economy through the fifteenth century. The core feature was the Levant trade (Braudel 1984: 132), so that Europe was still operating at the periphery of other world-economies. With the reorientation of Europe towards the Atlantic in the first half of the sixteenth century, the economic centre moves west and north and Antwerp becomes the new world-city. This was a city 'created by outside agency', as Braudel (ibid.: 145) describes it. It was the new crossroads of the world-economy for merchants engaged in both intra-European and transatlantic trades. Its predominance faltered in the 1560s and the locale of the world-city returned south, to Genoa. Braudel refers to this case as 'discrete rule' by merchant bankers (ibid.: 164) who operated 'by remote control' (ibid.: 159). Through financing the Spanish state, a small coterie of banker-financiers in Genoa was able to accrue the wealth of the Spanish Empire and dominate European finance. For Braudel, this was the last time a city without the backing of a strong state would be able to dominate the European world-economy. These 'city-centred economies' were always fragile affairs that could be undermined by the actions of state creditors (as happened to both Antwerp and, later, Genoa in the 1620s).

With the full establishment of the modern world-system in the sixteenth century, world-cities become directly subject to inter-state rivalry and appear as part of the rise of hegemonic states. There are just three examples of such unrivalled economic leaders in the history of the modern world-system, and each has created a new world-city: the Dutch Republic and Amsterdam in the seventeenth century, the British and London in the late eighteenth and nineteenth centuries, and the United States and New York in the twentieth century. Because the Dutch Republic was a relatively loose alliance of cities and provinces, Braudel treats Amsterdam's dominance as a halfway house between a city-dominated and a state-dominated world-economy. Nevertheless, because of its role within a hegemonic state, Lee and Pelizzon (1991) describe Amsterdam, along with London and New York, as a *hegemonic city*. This label for world-cities is most relevant in the realm of finance, where hegemonic cities outlast their hegemonic states in importance: Amsterdam remains the world financial centre through most of the eighteenth century long after the decline of the Dutch Republic; London's world financial prowess endures into the first half of the twentieth century, unlike British hegemony, which concludes about 1870; with New York in post-United States hegemony, the contemporary world city network is reached in which the city remains eminently important but in new 'global' circumstances, of which more later. Arrighi (1994) has developed Braudel's world-cities as international financial centres most comprehensively, adding Genoa to the three hegemonic cities in four *longue durée* cycles of capital accumulation.

Two features stand out in this world-cities argument. First, there is a continuity from a late medieval European world-economy to the twentieth-century modern world-economy. Braudel does distinguish between city-dominated and state-dominated world-economies but basically the same model operates. And second, this model is an extremely hierarchical one. A world-economy is seen as a process in which one city necessarily dominates; this is the essence of the model. Although recognized as complementary players in this economic game, all other cities are effectively ignored until it is the turn of a new city to lead. Both of these features will be critically scrutinized below.

National urban systems

Braudel's model of world-cities is useful for linking earlier world economies to the modern world-system but it does not give due weight to the remarkable eclipse of cities by modern states. It is not just that in the modern world-system his world-cities become part of a broader process related to hegemonic states and therefore can be deemed 'hegemonic cities'; the conversion from city-centred world-economies to a state-centred world economy affected all cities. The modern world-system ushered in a new political order based upon a new territorial organization. This 'Westphalia system' – named after the 1648 treaties that confirmed territorial states as the political building blocks of modern society – involved political centralization as states tried to eliminate or incorporate rival sources of power within their realms. City autonomies, 'privileges' enjoyed from the late medieval European world-economy, were among the victims of this restructuring of political power (Taylor 1995). Thus although the German Hanse survived into the seventeenth century, it was only a shadow of its former self, able to decline slowly because political centralization in Germany lagged behind that in the rest of Northern Europe. What had been the Hanse's core trading network in the Baltic was captured by merchants from Holland, where it became the 'mother trade' of the Dutch Republic, the springboard to world hegemonic success within the new inter-state system.

This early modern administrative centralization within states took political power away from cities; the later social centralization within modern states, nationalization, took away their city identities. The nationalist movements that flowered in the nineteenth century and blossomed in the twentieth century aspired to culturally homogenize populations within a given state's territory. This was the complete opposite of the cosmopolitan essence of cities in networks. The very idea of a large German presence in London with special privileges and a degree of self-government (the Hanse *Kontor*) is total anathema to the politics of nationalism. Cities became nationalized, mere components of nation-states, cogs in national economies. By the mid-twentieth century, instead of being viewed as nodes within great worldwide networks, major cities in the world were being seen in territorial terms: either as the capital city within a state, or as a 'regional centre' within a state. This territorializing of cities is perhaps starkest in Britain beyond London, where the great nineteenth-century industrial cities, the source of British world hegemony, famously become downgraded in status to mere 'provincial cities' in the twentieth century.

It is at this time that urban studies became a vibrant research field and, led by geographers and urban/regional planners, inter-city relations became an important focus of intellectual scrutiny. This required an answer to the basic question: how do cities relate to each other? The answer given was straightforward and pretty well unanimous: cities constitute national urban systems. The resulting research programme on national urban systems from the late 1950s to the early 1980s remains the most concerted effort to understand inter-city relations in the social sciences. The main themes of this research are delineated below because ideas from this programme have transferred over into world cities research. The reason is that key pioneers in the latter were schooled in, and sometimes had made important contributions to, the former.

Systems of cities

There is no doubt about how urban researchers conceptualized inter-city relations from the 1960s through to the 1980s: cities formed 'systems'. From Brian Berry's (1964) emblematic 'Cities as systems within systems of cities' and his 'integrated readings', with Frank Horton (1970), *Geographic Perspectives on Urban Systems*, through Larry Bourne's (1975) comparative *Urban Systems*, Alan Pred's (1997) definitive *City-Systems in Advanced Economies* and the basic reader *Systems of Cities* (Bourne and Simmons, 1978), to the more geographically specific Peter Hall and Dennis Hay's (1980) *Growth Centres in the European Urban System*, Stan Brunn and James Wheeler's (1980) *The American Metropolitan System* and Ron Johnston's (1982) *The American Urban System*, it is clearly stated that 'systems thinking' is necessary in order to understand inter-city relations. This choice of encompassing framework was highly significant: the idea of a system brings a huge intellectual baggage with it.

Bourne and Simmons (1978) discuss in detail the nature of urban systems. Starting with the basic definition of 'a set of interdependent cities comprising a region or nation' (p. 3), they argue that in order to understand how any one city changes – grows, stagnates or declines – it is not enough to study that particular city. The city should be seen as part of the larger system. For instance, the rapid growth of Chicago in the late nineteenth century can be understood only within the context of the westward expansion of the US urban system as a whole. But what kind of system is this? According to Bourne and Simmons, it is a complex system where the strong interaction between the elements (cities) creates 'feedback effects which regulate growth and change' (ibid.: 11). However, this 'complex system' was modelled as a relatively simple hierarchical order of cities: 'a national system dominated by metropolitan centres and characterized by a step-like hierarchy' (Bourne 1975: 13).

The hierarchical nature of the system was studied by investigating the distribution of city populations within a country. This was a chief means of confirming the 'system-ness' of sets of national cities. One particular model of city size distributions, the rank size rule, was employed to test out the system-ness. This model of the distribution of the populations of cities in a country specified a regular decline in size; in particular, when city rank is plotted against city population on logarithmic graph paper, the cities define a straight-line gradient. This was contrasted with 'primate city' distributions where the largest city in a country is far larger than all other cities, so there is no rank size gradient (Figure 1.4). The point of this contrast comes in the explanation. The rank size rule was interpreted as reflecting a condition of entropy in the urban system (Berry and Garrison 1958). This is reached when there are a myriad of forces acting on the cities in a country, so many in fact that they can be assumed to have random effects on city growth, which will thereby become proportionate to city size. This reformulation of the rank size rule as a statistical distribution allowed the model to be interpreted as the geographical outcome of a very complex national economy (like that of the United States). The primate distribution, on the other hand, is to be found in much simpler national economies (as in Third World countries) where there are only a few very strong

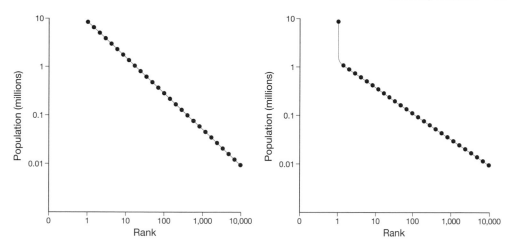

Figure 1.4 City size distributions

forces that favour the growth of just the leading city. In a seminal paper Berry (1961) transcended this duality by postulating a developmental model of city size distributions whereby primate distributions change into rank size distributions as economic development proceeds. As is the way with the development literature, this model assumed that all countries have the potential to develop fully fledged national urban systems, and this was duly predicted (Taaffe *et al.* 1963). These models continued to be used (e.g. Song and Zhang 2002). But the key point is that this specific form of hierarchy, the rank size rule, was deemed to be the result of the entropy in the system; hence it follows that cities in modern, industrial economies constitute national urban systems.

Practices of national urban systems analysis

What sort of research was carried out within this systems framework? Although a rich and large tapestry of studies was created – the two readers (Berry and Horton 1970; Bourne and Simmons 1978) brought to the fore nearly 100 papers – there was relatively little on the operation of the system itself. Such studies covered three main areas of research. First, the study of the evolution of particular national urban systems, of which the classic example is John Borchert's (1967) study of American metropolitan evolution, charted system change. In Borchert's study the system evolves through stages defined by the extra-systemic effects of new transport technologies: epochs defined by sail and wagon, railways and 'auto-air'. Second, there are studies of diffusion of innovations through urban systems. This was a hierarchical diffusion wherein new behaviours were to be found first in the larger cities, whence they were passed down to smaller cities. Linking innovations to growth potential, this research implied that 'major urban centres [were] capturing and redistributing urban growth' (Bourne 1975: 23). This leads to the third area of research: policy-orientated studies to regulate national urban systems. Such was Berry's goal for the US urban system: 'a unified and consistent national urban policy is clearly needed' (Berry and Meltzer 1967: 9). However, it was in other countries with greater acceptance of national and metropolitan planning instruments that such studies were most developed; Bourne's (1975) comparison of regulation policies for British, Swedish, Australian and Canadian urban systems represents the apogee of this approach.

One key feature that does stand out in reviewing this school of research is that despite the strong empirical tendency in the work, the evidential basis is usually quite limited. This is especially the case for actual relational measures between cities. Bourne and Simmons (1978), in their introduction to the part of their reader that deals with 'inter-city linkages, diffusion and conflict', complain that 'data inadequacies, conceptual vagueness and analytical difficulties have thus far permitted researchers to make only fragmentary descriptions and even fewer explanations of interurban interaction patterns' (p. 260). The result is that the section dealing with inter-city relations is actually the smallest of their seven sections of papers; it might be thought that this should be the most important section for a reader on systems of cities, but the editors provide only four papers on the topic out of a total of thirty-nine papers in the book. What kinds of data are described here? There are three basic types and they are representative of the whole literature. First, there are maps of infrastructure links between cities: Board *et al.* (1978) show maps of road traffic, airline traffic, rail freight, trunk telephone calls and labour migration to illustrate the structure of the South African national urban system. Second, there are diffusion maps indicating the date when a particular innovation was taken up across a set of cities: Pedersen (1978) shows such diffusion patterns both within a national urban system (Chile) and between such systems (in Latin America). Third, there are maps of the spatial structures of organizations in relation to the urban system: Pred (1978) shows how selected multi-locational firms are distributed across US cities. His findings are summed up by a phrase in the title of his paper, 'the complexity of metropolitan inter-dependence'. Since I consider Pred's work in this area to be pivotal to both this research school and the research reported in later chapters, his work requires some further scrutiny.

First the methodology: Pred (1978: 292) begins with the assumption that 'in an economically advanced system of cities large multilocational organizations are the major source of intermetropolitan and interurban interdependencies'. Thus they constitute 'the most appropriate objects to focus on when attempting to understand the interurban growth-transmission structure of systems of cities in post-industrial economies' (Pred 1977: 104). This focuses his research on 'intra-organizational linkages'. Firms are studied in terms of their geographies as expressed in the locations of administrative and production facilities. It is through this 'intra-organizationally based city system inter-dependence' that 'flows of administrative services, control, coordination and informa-tion' are transmitted. Thus these organizational links can be used to overcome 'the paucity of data' on interurban flows (ibid.: 126). However, they remain 'crude educated guesses' (ibid.: 127), surrogate measures of actual flows.

In terms of results, his method allows him to make statements about the 'control' of jobs in a given city; specifically, in which other cities are the headquarters of firms with plants in the given city? He makes the key point that hierarchical structures are to be found in the firms, *not* in the cities (Pred 1978: 297–8). This is part of his general concern that the city systems literature places 'excessive emphasis . . . on the hierarchical struc-ture of interdependencies' (Pred 1977: 17). Overall, his basic finding is that inter-city relations are much more complex than is usually appreciated (ibid.: 162). Instead of a simple city hierarchy he shows an 'unexpected' (Pred 1978: 307) criss-crossing pattern throughout the US urban hierarchy. It would seem that the dominance of the hierar-chical representation of inter-city relations consequent on the national urban systems model is misleading. It was shown that when this assumption was subjected to large-scale empirical investigation using relational data, the hierarchical coherence of the system becomes problematic. This is a complexity of a different kind to the theoretical concept of a 'complex system' and the reduction of that system to the rank size rule. Pred's is a messy empirical complexity, one that points to a fundamental revision of how to conceptualize inter-city relations.

There are two elements of Pred's work that I use later. First, I follow Pred's lead in solving the problem of inadequate inter-city data: I use information on the distribution of firms across cities. Although Pred collects his data within the context of urban systems, notice that this empirical approach is wholly consistent with the previous identification of an interlocking network described for Italian and German Hanse cities above. Merchant partnerships linked cities together in these city networks, just as multi-locational firms link cities together in more recent times. Second, Pred offers a challenge to the conventional view that inter-city relations are defined by a simple hierarchical order. Although distributions of city populations might suggest a hierarchy, long ago Lukermann (1966) pointed out that hierarchy can be shown only by using relational data. I can now add that on occasions when this is done, the urban hierarchy appears to be a quite problematic construct and thus will not feature as an operational assumption in my contemporary inter-city researches below.

Beyond the national?

As noted above, calling inter-city relations within a country a 'system' is a very strong statement that brings with it a huge intellectual baggage. Its origin within urban studies seems to have been the popularity of General Systems Theory in 1960s geography. Although this intrusion of highly structured thinking into geography was to attract early critics (Chisholm 1967), the framework remained for another decade or so. Such systems thinking was easily fused with contemporary planning ideas and thus became the necessary language for promoters of national urban planning: identifying 'feedback levers' created tools for 'system regulation'. But in what sense did the national sets of cities actually form 'national systems'?

This question was not widely asked. For instance, Bourne (1975: 14) simply asserts that 'the national urban system can be easily recognized'. Within systems thinking, the nearest researchers get to doubting the existence of national urban systems is when the openness of the system is discussed. If one eliminates the possibility of these being closed systems, the question of the degree of openness would seem to be of vital importance. But no: 'one neglected aspect of research on city systems is the effect of influences derived from outside the nation' (Bourne and Simmons 1978: 19). Bourne's (1975: 16) work is an exception and he identifies 'the international urban economy' as an extra-systemic influence. However, this is never clearly conceptualized and appears later in his discussion in the guise of the 'global (urban-economic) environment' (ibid.: 205) and the 'international urban/economic system' (ibid.: 206). When Simmons (1978: 62) directly broaches the subject he comes up only with the notion that national urban systems in smaller countries are more open (and therefore more unpredictable). In short, even when the question of openness is recognized, this does not lead to any serious questioning of the validity of the coherence of national systems thinking.

But in hindsight this ultra-territorial approach to cities seems quite odd. Were there actually more than a hundred separate national urban systems each operating behind their own boundaries? Take the example of Chicago's late nineteenth-century growth spurt that was mentioned above as being explicable as part of a national urban system expansion. This was surely not the whole story. Chicago was the centre of a hinterland providing commodities for Europe as well as the eastern United States. Was not its rapid growth consequent upon a supra-national trading pattern that encompassed a network of cities that straddled the Atlantic? Which brings up New York: number one in US applications of the rank size rule, might not its prime position be due to its traditional role as gateway between the United States and the rest of the world? If this is so, it is not enough merely to recognize the openness of the US national urban system; its leading city is dependent on connections beyond the 'system' and therefore problematizes the system's veracity. New York was and is part of a city network that transcends the national territory in which it is located. The same can be said of all major cities: their inter-relations are never respecters of boundaries. Twentieth-century London with its imperial and post-imperial links, its American and European links, could never be adequately understood as simply the biggest city in Britain. And the same goes for *all* primate cities: they appear to be primary only because of the unexamined research decision to study them within political boundaries. Their typicality within Third World states is not about the emergence of bounded systems but is all to do with 'external' linkage, the imperial and post-imperial relations of dependency that Gunder Frank (1969: 6) first taught us.

In many ways, the national urban systems literature operated as an intellectual paradigm to banish thoughts of major cities operating within trans-state networks. There are rare examples of recognition of the vital importance of the 'international' scale, notably Boulding's (1978: 155) prediction that the 'city's future is in the international system'. He even speculated on a 'new Hanseatic League' but conceded that this was only a 'pipe dream' (pp. 155–6). Other gazes into the future seem to have totally missed the subsequent 'globalization of cities'. The Brunn and Wheeler (1980) collection of essays is a case in point. For instance, when Phillips and Brunn (1980) update the technological steps of Borchert's (1967) model of the evolution of the US urban system, they do not use the contemporary literature on communication advances to speculate that the next step might produce inter-city relations at a new larger than national scale. Rather, they change the very basis for identifying changes and, reflecting the state of the US economy, they come up with the name 'slow growth epoch' (pp. 2–3). This reflected the contemporary national pessimism concerning the future of US cities and was even more explicitly reflected in Birdsall's (1980) discussion of the 'alternative prospects for America's urban future'. Even though the volume includes papers on urban growth related to science and technology (Malecki 1980), communications technology (Brooker-Gross 1980) and corporate control (Stephens and Holly 1980), all topics central to the later world cities literature, there are no references to how such processes might involve US cities in relations outside the country. As a containing paradigm, national urban systems research was certainly a good insulator against thinking seriously beyond the national.

Before I move on to consider contemporary world cities, mention must be made of one concept that sometimes appears in the urban systems literature: megalopolis. This concept was coined by Jean Gottmann (1961) to describe the pattern of urbanization linking US eastern seaboard cities (Boston–New York–Philadelphia–Baltimore–Washington, DC). Fitting uneasily into the national urban systems model, it is fleetingly mentioned in this literature (see Berry and Horton 1970; Johnston 1982: 126–7). This is a pity, since Gottmann's (1961) study is, to a large degree, about inter-city relations and he did consider transnational linkages with megalopolis acting as the 'hinge' between the US economy and the world. Here is a later description of megalopolis by Gottmann (1987: 2): 'Megalopolis is a chain of national and international crossroads . . . which owes its destiny to a web of far-flung and multiple networks of linkages with the whole world.' It would be nice to portray Gottmann's distinctive contribution to inter-city relations as a bridge between national urban systems writings and the world city literature, but to do so would be misleading. Although he made contributions to the latter before he died (Gottmann 1989), the reality is that his distinctive relational approach was seen as 'lacking precision as well as generality' (Berry and Horton 1970: 54) and was never seriously considered part of the national urban systems school.

World cities

As shown previously, Braudel (1984) used the term 'world-city' to describe leading cities in his world-economies. I have kept Braudel's hyphen to distinguish his historical use from the use of the term, without the hyphen, to describe the major contemporary cities that are the main concern in the remainder of this book. In fact, Braudel's concept is very narrow compared to all other conceptions of world city. This is so in two ways. First, he limits his identification to one such city at a time: as previously shown, they form a sequence. Second, the city's prowess is limited to financial control within the world-economy: this may involve just a small coterie of financiers in a city. Although Arrighi (1994) has found Braudel's model useful for understanding economic cycles through to the 'long twentieth century', such a restricted view of urban processes has little to offer for understanding contemporary world cities.

Gottmann (1989: 62) traces the term 'world city' (*Weltstadt*) back to Goethe, who applied it to both Rome and Paris at the turn of the eighteenth and nineteenth centuries. According to Gottmann, he was using the concept to identify the leading cultural centres of his world. However, usually the origins of the term are traced back to the work of Patrick Geddes, pioneer planner and general polymath. This also has a German connection since Geddes' (1924) contribution is to comment on a German proposal to set up a world league of cities (*Welt-Stadtebund*). Here the concept is presented as a broader idea of the city, as a planner's holistic subject. Peter Hall's (1966) *World Cities* takes its inspiration from Geddes and he describes six such centres: London, Paris, Randstad-Holland, Rhine-Ruhr, Moscow, New York and Tokyo. His is a comprehensive study of these leading cities, covering, as well as culture, politics, trade, communications infrastructure, finance, technology and universities. Within this range of topics Hall (1966) does indicate the critical role of economic functions to his world cities and as such presages most later writings on the subject. Being written during the heart of the era of national urban systems analysis, the study represents an antidote to such parochial national thinking, but, although well received, it did not precipitate a rival 'world' urban research agenda. The world city literature as a cumulative and collective intellectual enterprise begins only when the economic restructuring of the world-economy makes the idea of a mosaic of separate urban systems appear anachronistic and frankly irrelevant.

The downturn in the world-economy in the 1970s resulted in an economic restructuring that transcended individual states. Multinational corporations were identified as new major players on the world scene: in the words of the titles of two influential books of the time, these firms were found to have 'global reach' (Barnett and Muller 1974) and operated in a 'world without borders' (Brown 1973). Despite this terminology, the initial framework that developed for interpreting the restructuring was not globalization but the 'new international division of labour' (Frobel *et al.* 1979). Multinational firms were shown to be reorganizing parts of their production to take advantage of low Third World wages and, in addition, new links between computers and communications were beginning to make such worldwide production manageable. Whither 'national urban systems' in this changing world? Only cities in the United States, by far the largest 'national economy', seemed still to be recognizably interpretable as a national system (as will have been noted from the list of books that started the subsection 'Systems of cities' (p. 16), the last two (Brunn and Wheeler 1980; Johnston 1982) were only about the United States). New thinking about cities was required and a world cities literature emerged in the 1980s and flowered in the 1990s as a central theme of globalization.

To review this literature I employ the nugget approach. After some two decades, research on world cities has created a very large literature covering a wide range of topics and issues. I focus upon three key contributions to this new school of research that I identify as the important signposts *en route* to the world city network that is my subject. Other studies will be mentioned in passing, but no comprehensive review is intended here. In fact, I will be quite selective in my treatment of the chosen texts: I concentrate only on their contributions to understanding inter-city relations. My three selections, John Friedmann's (1986) 'world city hypothesis', Saskia Sassen's (1991) 'global cities' and Manuel Castells' (1996) 'space of flows' (in a network society), will not be a surprise to those familiar with the literature. Each one has been widely recognized as a seminal contribution to the development of the world cities literature; all three writers have responded to the demand for further reflections (Friedmann, 1995; Sassen, 2000; Castells, 2001).

Friedmann's world city hypothesis

The first major paper to link cities to the new international division of labour was by Cohen (1981), shortly followed by Friedmann and Wolff (1982), but the most influential paper has turned out to be Friedmann's (1986) statement of 'the world city hypothesis'. This self-styled succinct 'framework for research' (p. 69) has formed the initial benchmark of this school of research and has continued to be used as a spatial framework up to the present.

For Friedmann (1986: 69), the world city hypothesis is about 'the spatial organization of the new international division of labour'. This is described in terms of seven 'theses' that 'link urbanization processes to global economic forces' (ibid.). Only the first three theses are described here; they provide the foundation position, and the two key consequences for inter- and intra-city relations.

The *functional thesis* provides the fundamental premise of Friedmann's argument. It states that structural changes in cities depend upon the integration of a city into the world-economy and consequently upon the 'functions assigned to the city' (1986: 70). Three main functions are identified: headquarters functions, financial centres and 'articulator' cities that link a national or regional economy to the global economy. The very important cities will carry out all three functions. Thus world cities are seen as resulting from urban 'adaptation to changes that are externally induced' (ibid.). Endogenous conditions relating to particular cities, such as apartheid in South Africa, can modify the process but the overall pattern is an urban response to the new international division of labour.

The *hierarchical thesis* describes the inter-city relations in Friedmann's argument. It states that cities are the 'basing points' of capital, and the resulting linkages create a 'complex spatial hierarchy' (1986: 71). Cities are where corporations organize production and plan market strategies. The hierarchy is formed by taking into account a number of city characteristics: the importance of the city as a finance centre, corporate headquarters, international institutions, business services, manufacturing, transportation and population size. From this information Friedmann (1986: table 1) identifies two levels of hierarchy, which he terms primary (such as London) and secondary (such as Milan). These are then geographically arranged in two ways. First, there is a 'horizontal' division (north–south) defining core and semi-periphery cities showing nine primary cities in the former and only two in the latter (São Paulo and Singapore). Second, there are 'vertical' divisions (east–west) defining three continental subsystems: Asian, American and West European. The resulting schematic map entitled 'The hierarchy of world cities' has probably been the most potent pedagogic instrument in teaching and researching world cities (Figure 1.5).

The *global–local thesis* describes the intra-city relations in Friedmann's argument. It states that the city's role(s) in the world economy are directly reflected in the structure

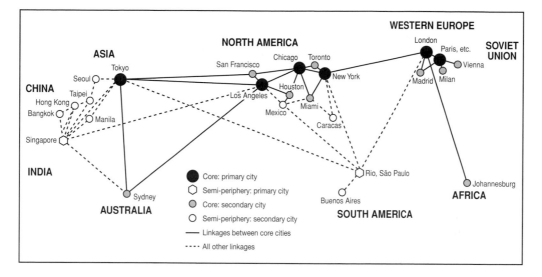

Figure 1.5 The world city hierarchy (from Friedmann 1986)

and change of the local economy. The control functions of cities mean that they are dominated by a number of expanding economic sectors relating to management, finance, corporate services and the media. This pattern has resulted in the massive growth of highly paid economic elites – corporate managers and specialist professionals – who provide a local market for a 'vast army of low-skilled workers' (1986: 74) servicing the new rich: hence the creation of a highly dichotomized labour force. The world city outcome is an extremely polarized social structure.

Thus are world cities firmly embedded into the world-economy. In his subsequent reprise of his theses, Friedmann (1995) confirms his position but with a slightly different emphasis. Spatial articulation is now emphasized. Cities are centres 'through which flow money, workers, information, commodities' and thereby they articulate the 'economic relations' of their 'surrounding "field" or region' into the global economy (p. 22). Therefore, in revising the world city hierarchy it is the scale of a city's spatial articulation that orders the cities (ibid.: 24, table 2.1): global (e.g. New York), multinational (e.g. Miami), national (e.g. Mexico City), and subnational/regional (e.g. San Francisco). But basically the same model obtains: he reaffirms a world city hierarchy.

What is to be learned about inter-city relations from Friedmann's seminal work? Certainly he put world cities and their inter-relations firmly on the research agenda of urban studies. Above all, he provided a new global vision of inter-city relations, a way of seeing connections between cities that transcend state boundaries. In short, his diagram of the world city hierarchy (Figure 1.5) became etched into the world city imagination. In his second paper Friedmann (1995: 26) makes it clear that these inter-city relations are 'a historically unprecedented phenomenon', thereby separating his model from the national urban systems research school to which he had been a major contributor (e.g. Friedmann 1978). Clearly, the transnational nature of the connections between cities represents a major break from the previous school, but it will have been noticed that the structure of the inter-city relations remains the same: hierarchical. In fact, his three continental 'distinct subsystems' (Friedmann 1986: 71) look very much like national urban hierarchies writ large, focusing on New York, London and Tokyo respectively. This is reinforced by the identification of only these cities at the global scale in his later 'articulations model' (Friedmann 1995: 24): using its financial prowess, each city articulates its 'subsystem' into the world economy. Thus I conclude that Friedmann's 'unprecedented phenomenon' certainly encompasses the scale of contemporary inter-city relations but not necessarily their structure.

Sassen's global trilogy

During the 1980s the conceptual context within which world cities were viewed changed: the new international division of labour gave way to a more encompassing language of the global. (This change can be seen in the two Friedmann papers described above.) Although Feagin and Smith (1987) provided a review of world cities from the perspective of the new international division of labour after Friedmann's (1986) seminal paper, the idea of a 'global capitalist city' appears earlier as a description of Los Angeles (Soja *et al.* 1983), and by the end of the decade global language dominated the literature: Knight and Gappert's (1989) *Cities in a Global Society* was soon followed by King's (1990) *Global Cities*. The latter identified the new world cities literature as 'a major paradigm shift in urban studies', resulting in 'global paradigms' (p. 3). However, it was in Knight and Gappert's collection of essays that the global framework was most emphatically stated. The editors argue that cities, having been 'eclipsed by nationalization', 'are now able to position themselves in a global society' (Knight and Gappert 1989: 15, 19). If one looks beyond economic relations in the 'emerging global context' (Knight 1989a: 40), 'global cities' are identified in a world where 'development is being driven more by globalization than nationalization' (Knight 1989b: 326, 327). Sassen (1991) encapsulated this new thinking in her detailed study of New York, London and Tokyo as global cities.

Contemporary globalization is premised upon new telecommunications and information technologies. The practical implementation of these innovations has incorporated two contrary yet complementary spatial tendencies: both decentralization and agglomeration of economic activities. This 'combination of spatial dispersal and global integration' is the 'point of departure' for Sassen (1991: 3) because it 'has created a new strategic role for major cities'. In other words, dispersal has created a demand for new control and organization functions that are found in cities. According to Sassen, the result is 'a new type of city' (ibid.: 4) that is essentially different from the historical banking and trading centres that I described earlier. She is explicit on this: global cities are 'nothing like' the cities in the Hanseatic League (ibid.: 4). Global cities function in four new ways (ibid.: 3) that form a 'virtual economic circle' exclusive to globalization (pp. 12–13): (i) the demand for control creates cities as 'command points'; (ii) this creates a demand for finance and business services, and cities become the 'key locations' for what become the leading economic sectors; (iii) as a result, cities become sites of production and innovation for these leading sectors; and (iv) cities constitute markets for the resulting leading sector production. Sassen identifies New York, London and Tokyo as the 'leading examples' of global cities and argues that, despite their very different provenances, each city has experienced the same economic processes with consequent social change (e.g. social polarization) in 'parallel' in a globalizing world economy.

In this way, Sassen (1991) follows earlier definitions of world cities (e.g. Friedmann's) but with a new emphasis on the production of financial and service products (p. 5). Although the theme of cities as international financial centres had been a major focus of research (Reed 1981; Thrift 1987) and was generally considered central to identification of world cities, Sassen's renewed focus on services is critical, as she integrates research on the latter (Stanback and Noyelle 1982; Daniels 1985; Wood 1987) into understanding global cities. Sassen's purpose is nothing less than 'seeking to displace the focus of attention from the familiar issues of the power of large corporations' (1991: 6). Global cities are more than simply 'command centres', they are the first 'global service centres'; in urban history.

What is to be learned about inter-city relations from Sassen (1991)? At first glance, as her book is essentially a comparative study of three cities with relatively little explicit discussion of the actual *relations* between the cities, it might be expected that there is little to use here. In actual fact, her emphasis on advanced producer services lies at the heart of how the world city network will be conceptualized in what follows. Her brief references to the 'vast multinational networks' of service firms and their 'global integration of affiliates' (p. 173) will be my starting point. However, hierarchy as the way cities relate continues to be implicit in most of her discussion. Such a hierarchical view of cities is to be found in how global cities relate to other cities in their respective states (chapter 6); couched in terms of sites of post-industrial versus industrial production (e.g. Detroit, Manchester, Osaka), it follows that globalization is accentuating 'national urban hierarchies'. Beyond individual states, most of the empirical evidence Sassen marshals on finance and services is used to create city rankings and thereby show that her chosen three cities are, in fact, the leading cities in a 'global hierarchy' of cities (ibid.: chapter 7), although the latter concept is problematized (ibid.: 169). The latter is important because, in noting the differences between the financial activities performed in the three cities, she goes on to ask whether 'the interactions among New York, London and Tokyo . . . suggest the possibility that they constitute a system' (ibid.: 168): 'What is emerging out of analysis of the multiplicity of financial and service markets concentrated in these cities is the possibility of a systemic connection other than competition – an urban system with global underpinnings' (ibid.: 169). Earlier, this 'system' was also described as 'the network of global cities' (p. 9), but this formulation is not developed; Sassen's (1991) emphasis remains overwhelmingly on ranking cities to illustrate that her group of three cities are located at the top of a global hierarchy.

Interestingly, in the new edition of *The Global City*, Sassen (2001) selects inter-city relations, described as 'the emerging transnational urban system', as one of three topics that are thoroughly revised (p. xvii). In a curious throwback to Friedmann (1986), Sassen reorganizes her 'global city model' (ibid.: xix) as seven hypotheses, one of which states:

> Fifth, these specialized service firms need to provide a global service which has meant a global network of affiliates or some other form of partnership. As a result we have seen a strengthening of cross-border city-to-city transactions and networks. At the limit this may well be the beginning of the formation of transnational urban systems.
>
> (ibid.: xxi).

Simply remove 'systems' and replace with 'network' and this is essentially the position upon which this book is based. The thorough revision occurs in chapter 7, whose title is changed from 'Elements of a global hierarchy' to 'Elements of a global urban system: networks and hierarchies'. However, although the chapter begins with a section entitled 'Towards networked systems', hierarchical relations continue to dominate the discussion. In this book I will attempt to avoid this 'hierarchy fetish'.

Castells' global spaces of flows

Although Castells is a renowned urban researcher, *The Rise of Network Society* (Castells 1996) is not intended as a contribution to the world cities literature. It is part of a trilogy of books that attempt to reformulate social studies for a global age. Castells argues that 'networks constitute the new social morphology of our societies' (p. 469) in the 'informational age'. The latter has replaced the 'industrial age', the context in which orthodox social studies has developed. However, with the advent of new enabling communication and information technologies, new scales, scopes and intensities of networking have emerged that are 'reshaping . . . the material basis of society' (p. 1). Having previously identified the existence of 'the informational city' (Castells 1989), in this later work he accepts Sassen's global city model and employs it to illustrate a key dimension in his 'social theory of space' in a new network society (1996: 378). One note before I start on this topic: key elements of Castells' spatial argument have remained relatively unchanged for over a decade; the description in Castells (1996: 378) is taken in large part from the earlier work (Castells 1989) and has not been modified in the later revision (Castells 2001).

Following Harvey (1990), Castells (1996: 411) defines space in terms of social practices. This means that the form that 'social' space takes is not universal; rather, it is historically specific. For Castells, space facilitates social practices that require the interaction of at least two people in a simultaneous combination of actions. For instance, the space that is a marketplace makes possible the social practice of trading by bringing together buyers and sellers on 'market day'. This example illustrates the traditional constitution of such spaces being defined by contiguity, the physical bringing together of interacting agents. But it is precisely this physical requirement that communication and information technologies are designed to overcome: in the informational age the simultaneity of time can be separated from the contiguity of space. Thus in network society the dominant form of space is no longer spaces of places, it is a new space of flows (ibid.: 412). In this form of space, places do not disappear but they become defined by their position within flows.

The space of flows is defined as a combination of layers of material supports for dominant social practices. Three are identified as being important, although discussion focuses on the second layer, suggesting a 'sandwich' structure. The first layer is the infrastructural support for social practices, flows that make non-contiguous simultaneity possible. Castells (1996: 412–13) defines this layer as grounded in a 'circuit of electronic impulses' based on communication and information devices. The relevant infrastructures based upon these technologies range from the global Internet to global airline networks. The third layer is constituted by the spatial organization of economic elites (ibid.: 415–16). These are flows that support the interests and practices of the 'technocratic–financial–managerial elite' (ibid.: 415). Examples include networks grounded in exclusive restaurants and cosmopolitan leisure complexes through to segregated residential and vacation locales.

The second layer is the meat in the sandwich, the space of the social practices that define society. It is constituted by agents who use the infrastructure networks to link together specific places to carry out 'well-defined' economic, cultural and political functions (Castells 1996: 413–15). The places of operation are termed nodes and hubs. Nodes are where strategically important functions take place and they link localities into the whole network. Communication hubs are places that function to co-ordinate interactions across the network. The functions of the network define places that become 'privileged nodes' (ibid.: 414). For instance, there are medical research networks that connect famous hospitals (e.g. in Rochester (NY) and Paris) as nodes co-ordinated by specialist medical conferences as hubs. Other examples of such worldwide flows are high-technology manufacture with multifarious intra-firm linkages from research and development centres through to assembly lines, and narcotics production and distribution with multifarious drug cartel linkages from peasant farming through to money-launder banking. All such networks are deemed to be hierarchical in nature (ibid.: 415). Global cities constitute just such a second-level network.

For Castells (1996: 415), 'the analysis of global cities provides the most direct illustration . . . of the space of flows in nodes and hubs'. The analysis he is referring to is Sassen's (1991) 'classic study' (Castells 1996: 379), and he quotes her argument for the uniqueness of the four new functions that define the global city (ibid.: 384). But he does extend Sassen's argument to a far larger number of cities so as to constitute a network. Sassen (1991) does not suggest that New York, London and Tokyo are the only global cities, but she does imply that these 'specific places' (p. 4) are few. However, using the work of Cappelin (1991) on interdependencies between medium-sized European cities, Castells argues that 'the global city phenomenon cannot be reduced to a few urban cores at the top of the hierarchy' (1996: 380). He postulates a 'global network' connecting centres 'with different intensity and at a different scale' whereby regional and local centres within countries become 'integrated at the global level' (ibid. 380). And in a second key break with Sassen, he argues that this 'spatial system of advanced service activities' (ibid.: 385) defines the global city as 'not a place, but a process' (ibid.: 386). Here he defines global cities as networked phenomena: whatever the particular status and roles of specific cities – the concern of Sassen – the really 'significant' feature is the network itself (ibid.: 385–6).

What is to be learned about inter-city relations from Castells? Although many of his ideas are derived directly from Sassen, Castells clearly provides a new context through which to view world cities: cities are part of a space of flows that in turn expresses the new network society. It is the resulting conversion of global city as advanced service centre into a global network of cities that is most significant for the remainder of this book.

A new spatial logic

Castells' idea of city as process has been used at the beginning and end of this chapter's substantive discussion of inter-city relations; specifically, to introduce the historical section and to conclude the world cities section. This is an indication of how important the idea is to this book. For Castells (1996: 386), it defines a 'new spatial logic' wherein the 'new' refers to the position held by all three of the authors featured above, namely that the emergence of world and/or global cities marks a qualitatively new phase in urban development. In fact, Sassen (2001: xix) avoids the term 'world city' precisely to distinguish her 'global city' from historically important cities that have been characterized as 'world cities'. By applying the idea to previous networks of cities I do not necessarily challenge this position – it is something to return to later – but I am suggesting that concepts such as space of flows and cities as networked entities are transferable across different historical specificities. Thus what I am basically taking from the above 'nuggets' of world city research is the necessity to think of cities relationally, as the product of networking activities. This is the meaning of new spatial logic employed here. In this short conclusion I provide a simple illustration of what such new thinking means for representing the world.

This is a book that promotes a city-centric view of the world and is written, in part, to counter the state-centric view of the world that emanates from most macro-level social science. I can now interpret these two positions as representing the world as different forms of space: cities in a network are a space of flows, whereas nation-states forming a territorial mosaic (the world political map) are a space of places. These alternative forms of space express different social processes: first, the nationalization of humanity over the past two hundred years; and second, the global challenge to this prioritization of places over flows today. Thus contemporary globalization can be viewed geographically as a tension between two 'world spaces': network and mosaic. As an illustration of the two forms of space existing both separately and together in contemporary social practice, the tension may be observed in the pages of *The Economist*, one of the world's leading business weekly publications.

Although *The Economist* is a UK publication, it operates as a truly global enterprise, with twenty editorial offices across the world in what is a roll-call of leading world cities: London, New York, Tokyo, Paris, Hong Kong, Frankfurt, Los Angeles, Brussels, Beijing, Washington, São Paulo, Berlin, San Francisco, Bangkok, Beirut, Delhi, Johannesburg, Mexico City, Moscow and Edinburgh. The last-named city provides the only hint of a British bias (especially given the omission of more important cities, such as Chicago, Singapore and Milan, from the list). Nevertheless, this is a magazine seemingly 'plugged in' to the world city network, located to report on the new 'globalizing' world. And yet the magazine remains dominantly territorial in its view of the world; it provides its readers with reports on regions and countries. Its text describes an international economy as a space of places: I refer to it as *The Economist* World I. However, an alternative picture can be found in the magazine between the pages of text: the advertisements describe a network world. They engage with a global economy as a space of flows; I refer to this as *The Economist* World II.

The Economist World I: the international economy as a space of places

The attraction of *The Economist* is to be found in the mixture of its reporting and analysis of current politics and business. It is not surprising, therefore, that a large chunk of its weekly contents is organized territorially. The world is divided into six regions: Britain, Europe, the United States, the Americas, Asia, and Africa and the Middle East. Of course, this is a very particular 'Anglo-American' view of the world, one in which the UK is not in Europe and the United States is not part of the Americas. No matter; the point here is the territorial frame. Remaining sections reporting on business and finance/economics also have many place-based stories, such as 'Germany's struggling retailers' and 'Japan's coalescing insurers' (both from the issue of 2 September 2000). However, the 'Indicators' section at the end of each issue best illustrates the mosaic nature of the magazine's view of the world.

Every week three pages of *The Economist* are allocated to statistics. There is a page each on 'Economic indicators' and 'Financial indicators', both of which feature all the main European economies (11), plus Australia, Canada, Japan and the United States. The third page, entitled 'Emerging-market indicators', provides information on the twenty-five largest economies in what used to be called the Second and Third Worlds. Attributes of these forty countries are described every week, thus creating a continuously updated world of containers, of national economies that together constitute an international economy. This is a world space of places where information on flows is largely missing. It is portrayed as *The Economist* World I in Figure 1.6.

The Economist World II: the global economy as a space of flows

Advertisers in *The Economist* are far less concerned with an international world. Browsing through their messages in editions of the magazine from May 2000 to January 2001, I notice that it is claimed that society is 'en route to a borderless world' (according to NTT DoCoMo) where there are 'communications without boundaries' (Avaya). These provide a 'global reach' (Global One) where there is a 'global market call for global

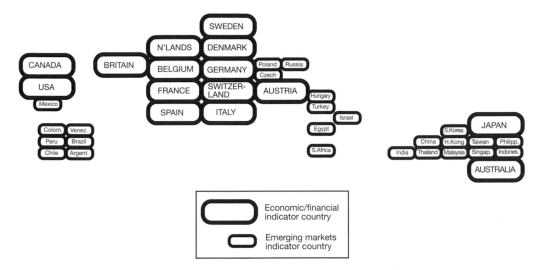

Figure 1.6 *The Economist* World I: the international economy as mosaic (from Taylor 2001b)

players' (Deutsche Post) who 'play to win on a global scale' (Cartesis). This is a *'new new economy'* (SAP), with 'a new opportunity, a new paradigm' (J. P. Morgan), a 'new world, new thinking' (IBM), which is 'unrelenting thinking' (Goldman Sachs) in a world where you 'never stop thinking' (Infineon). If you are not sure you are 'really up to speed for global change' (Commerzbank), you may wish to enrol in 'the only truly integrated, truly global Executive MBA' (Trium) or 'learn globally, study locally' (Fletcher School). But on the other hand there are 'answers to questions you don't even have yet' (quidnunc) and 'solutions to tomorrow's questions' (Dimension Data) readily available. This will enable people to 'change the world' (BT Openworld, PricewaterhouseCoopers) and 'bring order to chaos' (PricewaterhouseCoopers) so that the 'the possibilities are infinite' (Fijitsu) and 'unlimited' (Concert).

Behind all this optimism there is one basic message encapsulated by Morgan Stanley Dean Witter's bold slogan 'Network the World'. The promise is of a 'global network with seamless connectivity' (Concert) or 'seamless global network' (infonet), 'a network of 13,000 open minds' (UBS Warburg), 'networks of confidence' (Bull) and 'a network of excellence' (Goldman Sachs) and, with 'the new internet' (Nortel Networks), 'next generation networks' (telltabs) with 'global crossing' of access and information (Global Crossing) based upon 'more superhighway, less road rage' (Agilent Technologies). Among this invocation of virtual networks, there are less obvious but equally noteworthy references to more concrete networks such as 'the airline network for Earth' (Star Alliance). And behind all infrastructure networks there have to be people who make the system work. These people work in offices, and several companies proclaim their prowess in this respect: at one level 'the flexibility of 325 offices' (Maersk Sealand) and at another with 'over 5,000 offices world-wide' (HSBC). Such offices in their high tower blocks provide the familiar silhouettes of world cities and, more important, can be used to define the world city network.

Many advertisements mention cities – not just their headquarters location, but the particular network of cities in which the advertiser operates. I have focused upon just those advertisements that mention five or more cities for the nine months May 2000 to January 2001. There were forty-six such advertisements, which had a total 681 mentions of cities between them. In all, 154 networked cities were identified, with London, not

surprisingly, having the most mentions (appearing in thirty-nine of the forty-six advertisements). In Figure 1.7 just the cities that appear in five or more networks are shown.
These define a skeletal network of a global economy, the nodes of a world space of
flows. This is *The Economist* World II, the picture readers get from the magazine's
advertisements. It is also the space of flows that I investigate in detail below.

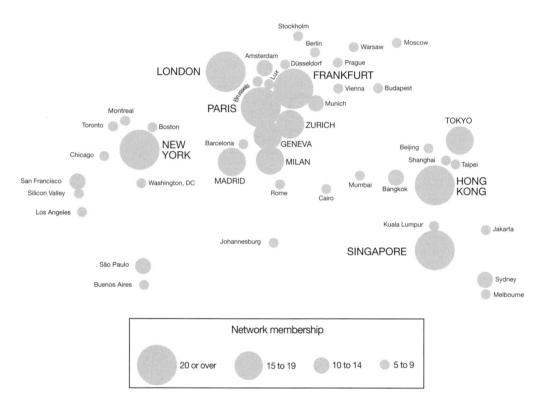

Figure 1.7 *The Economist* World II: cities as nodes in a global economy (from Taylor 2001b)

2 Back to basics

[I]t is impossible to tell the story of any individual city without understanding its connections to elsewhere. Cities are essentially open; they are meeting places, the focus of the geography of social relations.

(Massey *et al*. 1999b: 2)

In the previous chapter I present an uncritical, selected review of the world cities literature. In fact, of course, this literature is highly contested in several respects – see, for instance, the critical discussions by Soja (2000: chapter 7) and Smith (2001: chapter 3). The starting point of this chapter is a particular critique that I believe to be absolutely fundamental to any further advances in the understanding of world cities. My critique concerns the world city literature's evidential basis. In particular, I argue that there is an acute empirical deficit for studying the external relations of cities. As I have shown, there are numerous statements in the literature about inter-city relations at the global scale but when these are investigated they are often based upon the flimsiest of empirical evidence. This is a quite remarkable state of affairs given the supposed nature of world cities as components of global hierarchies and networks. The first section of the chapter illustrates the extent of the problem; the rest of the chapter reflects on the implications of this situation.

The key question is what to do about it. The natural response of a researcher to such a situation is to hunt down information and to try to create the data that are needed. This is an understandable reaction but it is not the right one in this case. The point is that this cannot be just an empirical problem; it indicates a serious theoretical limitation in thinking. Data are not something that can be created separately from ideas about how cities operate. There is an inherent iteration between theoretical and empirical phases of research, and in this chapter I argue that before beginning to measure relations between cities, the objective of Part II, it is necessary to rethink how cities operate under conditions of contemporary globalization. I take the view that this is so serious that a fresh start for studying inter-city relations is required. This is the 'back to basics' of the chapter title.

To find some alternative basic ideas on inter-city relations I have searched out a classic study of urban change that lies outside the national urban systems tradition. Jane Jacobs' (1984) *Cities and the Wealth of Nations*, written nearly two decades ago, takes as its starting point what cities have done historically – in particular, projections of their economic influence over space. Without having national boundaries as obstacles to thinking relationally, Jacobs avoids the ubiquitous world cities position of always seeing cities in hierarchies. This hierarchical presumption is the bane that world cities research inherited from the urban systems tradition: it has survived into globalization studies precisely because it has never been confronted with adequate evidence to test its veracity. When hierarchy appears in later chapters it is treated as an empirical question to be decided upon using evidence.

The main argument of this chapter takes the form of a presentation of Jacobs' ideas on inter-city relations, tailored to inform an understanding of cities operating under conditions of contemporary globalization. To help me in this task I use the world-systems analysis of Immanual Wallerstein (1979). His basic core–periphery model appears in the previous chapter as the spatial framework in which Friedmann set his world city hierarchy; here I explicate that frame further. As social science modes of study, Jacobsean economics and world-systems analysis are very different in many respects but they do share a rare ontological heresy: both Jacobs and Wallerstein eschew the orthodox view that societies and economies are politically bounded by territorial states. They agree that studying, say, the 'American economy' or the 'French economy' is not a meaningful starting point for understanding social change in those countries. However, the similarity quickly evaporates when they propose their respective alternatives to this orthodoxy: Jacobs goes below the state to study city economies, Wallerstein moves upwards to the world-economy. Combining these ideas, I create a 'double move' in geographical scale of analysis. This move is deemed necessary because the challenge to understanding cities posed by processes of contemporary globalization has exposed an ontological shallowness within social science that contemporary urban studies has only partially begun to rectify. This chapter is my contribution to putting this right so as to view cities without state-focused lenses.

The argument divides into four sections. After the first section on the critical empirical deficit, the second section is devoted to explicating Jacobs' arguments about the economic nature of cities. 'Dynamic city' regions are defined as the basic entities of economic change. A pure economic landscape of city regions is constructed and I use it to respecify the global core–periphery model. In the third section the argument is developed by bringing the modern state back in as an addendum that changes this abstract landscape to a political economy world that more closely resembles the real world that cities exist in. However, this does not entail a return to studying the 'national economy' since this concept is treated as a myth, a construct that forces economic processes into political containers. Whither the 'national urban hierarchy' – this is shown to be a political economy artefact. In a short concluding section, world cities are interpreted as Jacobsean 'dynamic cities' to provide a new starting point for conceptualizing, measuring and analysing the world city network.

The evidential crisis in world city research

The world city literature has attracted many sceptics. Kevin Cox (1997: 1), for instance, refers to 'so-called "world cities"', providing a salutary reminder of the dangers of attractive theoretical ideas turning into taken-for-granted conceptions. And the sceptics have had much going for them: the world cities literature is vulnerable to criticism for its dearth of evidence backing up its propositions. John Short *et al.* (1996) have identified this as the 'dirty little secret' of world cities research, and elsewhere I have referred to it as the 'data deficiency problem' (Taylor 1997) and the Achilles heel of the literature (Beaverstock *et al.* 2000a). The problem largely concerns information on relations between cities. In fact, disquiet concerning the paucity of evidence behind Friedmann's 'world city hypothesis' surfaced in the initial published discussion of the paper (Korff 1987) and it might have been thought that this problem should have been sorted out by now. But no: a brief exploration of what the literature says about US cities in globalization confirms the continuing nature of the problem.

Basically, a review of the US literature shows a focus on either comparative studies or individual case studies. Of course, Sassen's (1991) classic study of 'global cities' begins the 'comparative tradition' by comparing New York to London and Tokyo. As shown previously, in marshalling her evidence, she makes a major contribution to comparative studies but says little on the relations between these cities. The same can be said for Abu-Lughod's (1999) study of New York, Los Angeles and Chicago as 'America's world cities'. Other important comparative studies are by Farnstein *et al.* (1992), Gordon and Richardson (1998), Kresl and Gappert (1995), Kresl and Singh (1999) and Rondinelli *et al.* (1998). In addition, there is a rich literature on individual US cities as world cities (e.g. for New York: Crahan and Vourvoulias-Bush (1997), Logan (2000), Markusen and Gwiasda (1994), Mollenkopf and Castells (1991), Scanlon (1989), Shefter (1993); for Miami: Hansen (2000), Nijman (1996, 1997), Sassen (1993); for Los Angeles: Dear (2002), Scott and Soja (1998), Suarez-Villa (2000); for Washington, DC: Abbott (1996, 1999), Fuller (1989), Stough (2000); for other US cities: Abbott (1993), Esparaza and Krmenec (1994), Ganz and Konga (1989), Harvey (1996), Hicks and Nivin (1996), Walton (1996)). Comparative city studies and individual case studies are vitally important for understanding US cities in globalization (Knox 1996), but they cannot be a substitute for researching how US cities are connected to each other and to other world cities. In contrast, US inter-city studies are relatively rare, and usually cover only Castells' (1996) first level of the space of flows, infrastructural patterns (Malecki 2002; Sassen 2002), both actual (e.g. airline networks: Smith and Timberlake 2002) and virtual (e.g. Internet: Townsend 2001; Zook 2001). Although such studies provide specific important insights into how US cities are linked within global networks, they do not measure actual inter-city social practices that have created a world city network. A few other researchers have studied inter-city relations through questionnaires (Esparaza and Krmenec 2000) and official trade data (Drennan 1992), but they deal with just a small selection of US cities. Lyons and Salmon (1995) cover more cities but consider only the corporate headquarters locations in US cities. In short, to look for any general empirical understanding of US inter-city relations in the literature is to look in vain. And the same can be said for world city studies in other countries and world regions. This is a great paradox of this literature: there is a dearth of research on the connections between world cities and yet the latter cities' pre-eminence is based precisely upon those under-researched connections!

This is a very serious indictment of the world city literature and requires careful specification. I detail the problem in two ways. First, I look briefly at some case studies of influential writings on world cities that use lamentable empirical backing for their arguments. These are presented as serious blemishes in otherwise sophisticated arguments. Second, I report the results from a more comprehensive survey of the evidential basis in key texts in the world cities literature. This shows a systemic weakness in the world cities literature.

Theoretical sophistication and empirical poverty

Since Friedmann's (1986) world city hierarchy (Figure 1.5) is the starting point for so much discussion about world cities, this is where I will begin in my evidential critique of the literature. According to Abu-Lughod (1989: 32), the origin of this mapping was 'a base map provided by Japanese Airlines'. No doubt this explains why the three cities with most connections in Figure 1.5 are in the Pacific Rim (Tokyo, Singapore and Los Angeles) and not in the North Atlantic region (e.g. New York and London). In his later, more sophisticated discussion of 'a hierarchy of spatial articulations', Friedmann (1995: 23) laments the 'lack of unambiguous criteria for assigning particular cities to a specific place in the global system', so that 'establishing such a hierarchy once and for all may . . . be a futile undertaking'. Because of the volatility of the world economy, Friedmann goes on to argue that 'a rough notion' of rank 'without further specification' is all that is needed. This seems to me to be something of an empirical cop-out. My view is that it is better to provide precise specification of inter-city relations for a given time so that criteria are no longer unambiguous. Only in this way can the presumed global urban volatility be confirmed and changes measured.

The most informed creation of a hierarchical global structure of cities is that of Roberto Camagni (1993). His diagram of 'the hierarchy of city-networks' (Figure 2.1) is well cited in the world cities literature. The argument is theoretically sophisticated because he combines three 'logics of spatial organization' – territorial (state), competitive (hierarchical) and network (co-operation) – into a single argument. However, he begins his discussion on 'towards a new theorization' with the following ominous statement: 'In spite of actual weaknesses in the empirical inspection, the theoretical research programme appears challenging and worthwhile.' The result is a hierarchy consisting of three layers of city networks with 'world cities' perched firmly at the top, 'national cities' in the middle and 'regional cities' at the bottom. For all the sophistication of the theoretical argument, this looks very much as though a global level of inter-city relations has just been added to existing lower levels. Without any empirical evidence to suggest otherwise, this may be a reasonable assumption, the best guess of contemporary inter-city relations that can be derived from the national urban system tradition. However, there is a serious problem with this model: it reifies geographical scales. That is to say, it treats scales as real entities rather than intellectual devices for ordering the world. Like other examples of such reification (Taylor 1982), the scales identified actually reflect the orthodox state-centric world model featuring three scales pivoting about the state as the 'connecting' (middle) scale level. Of course, the world does not consist of separated geographical scales.

These two examples of empirically challenged arguments are by leading world cities researchers attempting to produce frameworks for understanding their subject. If these primary researchers have difficulties in providing convincing accounts of inter-city relations at a global scale, it cannot be expected that secondary researchers – users of world city research – will fare very well as they try to integrate world city ideas into their own theoretical positions. I provide three examples of world city literature users who are let down by the empirical deficit.

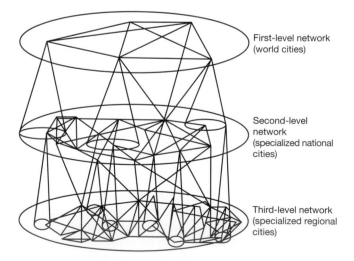

First-level network (world cities)

Second-level network (specialized national cities)

Third-level network (specialized regional cities)

Figure 2.1 Nested hierarchies (from Camagni 1993)

I begin with a book I have already considered as a major contribution to understanding world cities: Manuel Castells' (1996) *The Rise of Network Society*, whose theoretical concept of 'space of flows' is undoubtedly one of the most influential ideas in the field, and rightly so. Castells (1996: 26) is explicit that new data generation is not part of his brief and therefore he has to rely on data as published by other researchers. The result when illustrating his contemporary space of flows is so incredibly broad grained as to be derisory. The only evidence he provides on flows through nodes and hubs is a diagram based upon a set of worldwide information from Federal Express, originally analysed by Michelson and Wheeler (1994). I have no need to reproduce his illustration of this pattern of flows (Castells 1996: 383) because it is so simple. I can just list all the nodes: there is one origin, the United States, with flows to nine destinations comprising six cities – Hong Kong, Tokyo, Sydney, Mexico City, London and Brussels – and three places – Canada, South America and Puerto Rico. Thus, while I go along with Castells' conceptualization of a global space of flows, I must note that the evidence he marshals is mightily unimpressive.

Peter Dicken's (1998) *Global Shift* is the most influential geographical book on contemporary globalization. The value of this key textbook lies to a large extent in its bringing together reams of evidence to describe contemporary transformations in the world economy. But not so with world cities; here he provides minimal evidence. This consists of a diagram that is intended 'to give an impression of a connected network of cities' (p. 209) because 'the links shown are diagrammatic only'. He bases his diagram on Friedmann's (1986) 'world city hierarchy' (Figure 1.5) but with some additions to US and West European 'subregions'. In Figure 2.2 I have abstracted the European part of this network to show some of the very odd linkages that are displayed by Dicken. For instance, the route from Düsseldorf to London goes first to Brussels and then on to Paris before finally reaching its destination. Why there should be such a three-step connection in this electronic communication age is not explained; it is certainly not

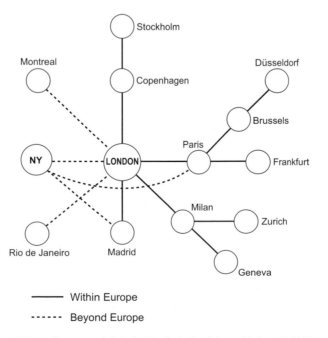

Figure 2.2 Linkage of West European 'global cities', derived from Dicken (1998)

based upon any actual evidence that any such flows take place. But remember, these are only 'diagrammatic links', again mightily unimpressive fare for an otherwise impressive text packed full of relevant information on other globalization topics.

The Open University texts in the Understanding Cities series, multi-authored with a consistently high level of theoretical sophistication, are the best textbooks available for contemporary urban studies. However, the systematic marshalling of empirical evidence is not one of the fortes of this impressive set of authors. Concern for relations between cities is a key feature that makes these books attractive, but this does not mean they escape from the endemic dearth of evidence on this subject. For instance, Massey (1999) in her chapter 'Cities in the world' begins with a section entitled 'Cities interlinked' that sounds promising but hardly lives up to its potential. It uses the fascinating example of the Square of the Three Cultures in Mexico City to illustrate changing locations of the city in historically different networks. On one side of the square there is the Aztec pyramid which for Massey represents the linking of the pre-conquest city into regional trade networks (illustrated as six (diagrammatic) flows on p. 106). On the second side the Catholic cathedral represents the transformation of the city to a colonial centre dependent upon transcontinental flows (shown diagrammatically as an arrow from Spain, p. 107). On the third side modern buildings represent independent Mexico's current industrializing project to become 'modern'. For some reason there is no diagram of flows for this complex of buildings, even at the level of purely diagrammatic links. (In fact, the state's foreign affairs department is housed here, so that there are numerous political flows to and from Mexican embassies across the world.) Thus the example makes one point very clearly – that city connections change – but completely neglects the present! Another chapter with a promising title – 'Cities of connection and disconnection' – by Amin and Graham (1999) does focus on the present and again encourages thinking 'relationally' about cities, but the only direct evidence that is provided takes the form of a 'virtual single office' linking together just six cities (p. 11). An important chapter by Allen (1999) provides the most thorough recent discussion of power among world cities, but he backs up his argument with two figures (pp. 193, 201) that are simply based upon Friedmann's (1986) hierarchy (Figure 1.5). The bizarre outcome is that this recent discussion of power excludes some of the most rapidly globalizing cities of the 1990s – Moscow, Beijing, Shanghai – simply because it uses a 'world city hierarchy' (purportedly) describing the situation before the end of the Cold War. Once again, with respect to quantity of evidence, these books are mightily unimpressive.

These 'evidential blemishes' in otherwise excellent books are symptomatic of a crisis in the world city literature. This warrants systematic evaluation before I proceed any further. To achieve this end requires the construction of what I term an evidential structure of the literature.

The evidential structure of the world cities literature

By the evidential structure of a literature I mean the nature of the empirical material that is marshalled to describe and test the propositions of a research field. It involves selecting key texts in an area and investigating the data employed with a view to finding particular biases in the resulting patterns. I have argued previously (Taylor 1997) that this is indeed the case for the world cities literature in two particular aspects. First, statistics in general have their origin in servicing the information needs of states, and this has resulted in the contemporary world being measured through state-centric data. Thus sometimes discussions of cities are illustrated only with data about states: for instance, Mayerhofer and Wolfmayr-Schnitzer's (1997) study of gateway cities is almost entirely based upon national data and Andersson's (2000b) description of 'financial gateways' uses only state-level data. Second, statistics in general have developed a critical bias towards measuring attributes at the expense of connections. Although some flows – of people, of commodities, of information – are measured, they pale in significance when compared to the quantity and quality of attribute data. Both these features of statistics have the potential to be quite debilitating for the study of inter-city relations.

For studying the evidential structure of the world cities literature I have chosen six key texts which will feature in the readings of all who research and teach the subject matter. The two key texts by Saskia Sassen (1991, 1994) need no justification. Similarly, there are three collections of essays that are obvious candidates for this exercise (Brotchie *et al*. 1995; Knox and Taylor 1995; Lo and Yeung, 1998); these have the advantage of adding a variety of authors to the survey. The final text focuses on the enabling potential of electronic communications, which is the key technology connecting world cities: Graham and Marvin (1996) deal explicitly with this technology in relation to the contemporary city. I believe these six books provide a reasonable cross-section of the world cities literature in the 1990s.

For each book I have concentrated upon the tables and figures used to inform the written text. Eliminating tables and figures that are not based upon specific data (such as a schematic diagram) or that are not place-specific (such as comparing the efficiencies of different technologies) leaves 426 pieces of evidence for analysis. Each table or figure was first classified as to whether the evidence used was state based or city based. Where both cities and countries appear in a table or figure a subjective decision was made in terms of which of the entities the table or figure most informs. Second, the evidence was classified in terms of whether the data concerned attributes or were relational. For the latter there has to be an origin and a destination. For instance, a table ranking countries by foreign direct investment is deemed attribute; if the locations from where investments came are given it becomes relational. In addition, the subject matter of tables and figures was recorded to aid in interpreting the results.

The resulting evidential patterns are shown in Table 2.1. The first point to make is that the texts are very empirically based: the presence of 426 pieces of empirical evidence in just six books does suggest that the authors have not felt themselves to be intimidated by the potential data problems alluded to earlier. However, the resulting patterns are not what you might expect from data with which to study world cities. Certainly there are more tables and figures on cities (263) than on states (163), but not

Table 2.1 *An evidential structure of world city literature*

Evidence	Source: A	B	C	D	E	F	All
SUBJECT							
Cities	46	16	17	102	16	66	263
States	30	17	3	51	12	50	163
Ratio	1.53	0.94	5.67	2.00	1.33	1.32	1.61
TYPE							
Relational	10	4	5	11	10	11	51
Attribute	66	29	15	142	18	105	375
Ratio	0.15	0.14	0.33	0.08	0.56	0.10	0.14

Key to sources: A = Sassen (1991); B = Sassen (1994); C = Knox and Taylor (1995); D = Brotchie *et al*. (1995); E = Graham and Marvin (1996); F = Lo and Yeung (1998)

by as much as might be expected since the latter are the explicit subject matter for all six books. In fact, much of the surplus of city evidence over state evidence is accounted for by the fifty-three tables and maps that feature simple city population figures from state censuses, hardly very sophisticated evidence of world city processes. Of course, there is nothing intrinsically wrong with having information on states in writings on cities: relating processes to different geographical scales is to be applauded. It is the balance of evidence that is at issue here. Obviously it can be argued that states provide the context within which the cities are changing and therefore tables on state trends are necessary. Well, yes again, but it is worthwhile putting the matter the other way round to get a clearer perspective. Cities and their regions constitute countries and national economies are primarily the sum of activities carried out in towns and cities. But it is not expected, and certainly it is not found, that social science texts on states and national economies have about one-third of their evidence on cities. Perhaps they should, but that is a different question; the point here is the relative quantity of evidence about states in a literature about cities.

Of course, the cities in this literature are not just any cities; they are world cities, with the presumption of incorporating activities that transcend the particular states they are located within. To understand this requires information on connections. Here I find a massive weakness in the evidential pattern of the world city literature, with attribute data (375) far exceeding relational data (51). Furthermore, it should not be assumed that the relational data all relate to cities: this is true of just thirty-one tables and figures. Think about this: a number of only thirty-one pieces of evidence on such a key matter – cities and connections – is astoundingly low. In a literature that abounds with concepts such as city networks and global urban hierarchies, only 6 per cent of the evidence directly informs such central notions! When I turn to a more detailed look at the figures for individual texts and what they mean, even this meagre grand total gives a rosy picture of the situation.

If we look at the detailed figures for individual texts, there are not many surprises given the above grand totals. The highest ratio for relational data is for Graham and Marvin's (1996) work on telecommunications, but even here attribute data still dominate. Furthermore, half of their relational data are about intra-city connections, so those data do not inform inter-city relations. In fact, many of the data on connections between cities are from analyses of airline flights, where information is easily accessible for cities as represented by their airports (e.g. Keeling 1995; Rimmer 1998). In Rimmer's (1998) work he contrasts the difference between telecommunications and air passenger networks. For the former he argues that '[m]any of the propositions on telecommunications and world cities have be taken on trust' (p. 451) and illustrates connections with a diagram of flows between countries. For air passenger networks he is able to produce diagrams that feature cities. However, it should be noted that airline flows are in any case problematic measures of relations between world cities. In a map of European airports presented by Kunzmann (1998: 49), after the big three (London, Paris, Frankfurt) there are fourteen second-order airports identified, such as Munich, Milan, Madrid and . . . Palma (Majorca). The last is, of course, the most popular holiday destination in Europe, hence its high ranking among important world cities. Airline passenger flows are constituted by many processes outside world city processes and they will all be reflected in gross figures – for Miami, for instance, vacation, retirement and commercial traffic are all mixed together. Finally, it should be noted that the airline traffic analyses in these books have all incorporated an insidious form of state-centric bias. All base their analyses on 'international passengers', which results in a downgrading of US world cities in particular. The classic example is Chicago – it appears on only one of Rimmer's maps (p. 460), courtesy of 'fourth-level' linkage to Toronto, while Dublin, for instance, appears on all maps because of its 'first-level' link with London. Nobody would want to argue, of course, that Dublin is more important than Chicago as an airport hub in the world-economy; it only appears that way when reliance is placed upon international data.

Finally, before I leave the question of evidence, one particular point needs to be emphasized. It might be thought that because there is information on cities that allows them to be ranked in importance, this provides evidence that they constitute a hierarchy. But no; the information merely shows that cities vary across a given measure, it says nothing about relations between the ranked cities; ranking is not the same as showing a hierarchy. Lukermann's (1966) early intervention cannot be emphasized enough: to repeat, defining a hierarchy requires more than producing attribute measures. For a hierarchy to exist there has to be some notion of control up and down different levels: 'each hierarchical level has autonomy over orders below itself, while being dependent on those above' (p. 17). Thus when Sassen (1991), in her chapter called 'Elements of a global hierarchy', uses rankings of cities (pp. 170, 176, 181), she is not showing that a global city hierarchy exists. This is because there has to be evidence of 'a *line* of command' (Lukermann 1966: 18, Lukermann's emphasis). To rank is merely to order by a size measure; it need have no relation to hierarchical structure (Taylor 1997).

In conclusion, I think the information I have presented on the evidential patterns of the world cities literature does suggest a severe structural problem due to the state-centric biases in data available to social science researchers. The evidence for asserting that world cities do in fact constitute nodes in a network of flows within the world-economy is very patchy indeed. Since the concept of world city has no meaning without connections, it seems reasonable to suggest that this literature has been suffering from an evidential crisis.

Cities in globalization

One clear effect of the evidential crisis has been the failure for there to emerge any agreement on just which cities are world or global cities and which fail to qualify. Table 2.2 shows sixteen 'rosters' of world cities, global cities and international financial centres from different sources. There are only four cities all authorities agree upon: London, New York, Paris and Tokyo. And yet there are seventy-eight other cities in the table that at least one source names in its roster. In general, this reflects the failure of the world city literature to provide specifications of the city concepts that are used. Thus there is no way to define a 'cut-off point' to identify which cities do not qualify for inclusion. This is found right at the beginning of the literature, where Friedmann (1986: table 1) states that 'in principle it would have been possible to add third- and even fourth-order cities to our global hierarchy'. Which are these 'other' world cities? Also, although Sassen (1991) is clear that New York, London and Tokyo do not exhaust the class of global city, mention of other cases in the text is casual, with no suggestion that the universe of this new type of city is known. But, in any case, is the idea of a given roster of world or global cities a good one?

You do not have to be a technology determinist to appreciate the importance of electronic communications to the rise of contemporary globalization. Doing business in different parts of the world will require operating through cities for advanced producer servicing, as Sassen emphasizes, but the consequent links and connections do not have to have a 'tree' structure. In developing a business project in Manchester that involves, say, working with people in Brisbane, there is no need to route the inter-city connection via London at one end and via Sydney at the other. Messages involving information, knowledge, strategy, planning, instruction, etc. can be and are directly relayed between

Table 2.2 *Cities cited in world city research*

City	Source: F1	F2	PK	DK	SS	FG	NO	RP	HY	CO	TH	KA	LP	RE	SK	GW
Amsterdam	o	x	x	x	x	x	x	x	o	o	o	o	x	x	x	x
Atlanta	o	o	x	o	o	x	o	o	o	o	o	o	o	o	o	x
Bahrain	o	o	o	o	o	o	o	o	o	o	o	o	o	x	o	o
Bangkok	x	o	o	x	o	o	o	o	o	o	o	x	o	o	o	x
Barcelona	o	x	o	o	o	o	o	x	o	o	o	o	o	o	o	x
Basle	o	o	o	o	o	o	o	o	o	o	o	o	o	x	o	o
Beijing	o	o	o	o	o	x	o	o	x	o	o	o	o	o	x	x
Berlin	o	o	o	o	o	o	o	o	o	o	o	o	x	o	o	x
Bonn	o	o	o	o	o	o	o	o	x	o	o	o	x	o	o	o
Boston	o	x	o	o	o	o	o	o	o	o	o	o	o	o	x	x
Brussels	x	o	x	o	o	x	x	o	o	o	o	x	x	x	x	x
Budapest	o	o	o	o	o	o	o	o	o	o	o	o	o	o	o	x
Buenos Aires	x	o	o	x	o	o	o	o	o	o	o	x	o	x	o	x
Cairo	o	o	o	x	o	o	o	o	o	o	o	o	o	o	o	o
Cape Town	o	o	o	o	o	o	o	x	o	o	o	o	o	o	o	o
Caracas	x	o	o	o	o	o	o	o	o	o	o	x	o	o	o	x
Charlotte	o	o	o	o	o	o	o	o	o	o	o	o	o	o	x	o
Chicago	x	x	x	o	o	x	x	x	o	x	x	x	x	x	x	x
Cologne	o	x	x	o	o	o	o	o	o	x	o	o	o	o	o	o
Copenhagen	o	o	o	o	o	o	o	x	o	o	o	o	x	o	o	x
Dallas	o	o	x	o	o	o	o	o	o	o	x	o	o	o	o	x
Detroit	o	o	o	o	o	o	o	o	o	o	o	o	o	o	x	o
Düsseldorf	o	x	x	o	o	o	x	x	o	x	o	o	o	x	x	x
Frankfurt	x	x	x	x	x	x	x	x	o	x	o	x	x	x	x	x
Geneva	o	o	o	o	o	o	x	x	o	o	o	o	o	o	o	x
The Hague	o	o	o	o	o	o	o	o	o	o	o	o	o	o	x	o
Hamburg	o	o	x	o	o	o	x	o	o	o	o	o	o	x	o	x
Hartford	o	o	x	o	o	o	o	o	o	o	o	o	o	o	o	o
Hong Kong	x	x	o	x	x	x	x	x	o	o	x	x	x	x	o	x
Honolulu	o	o	o	o	o	o	o	o	o	o	x	o	o	o	o	o
Houston	x	x	x	o	o	o	x	o	o	o	o	x	o	o	o	x
Istanbul	o	o	o	o	o	o	o	x	o	o	o	o	o	o	o	x
Jakarta	o	o	o	o	o	o	o	x	o	o	o	o	o	o	o	x
Johannesburg	x	o	o	x	o	x	o	x	o	o	o	x	o	o	o	x
Kobe	o	o	o	o	o	o	o	o	o	o	o	o	o	x	o	o
Kuala Lumpur	o	o	o	o	o	x	o	x	o	o	o	o	o	o	o	x
Lisbon	o	o	o	o	o	o	o	o	o	o	o	o	x	o	o	o
London	x	x	x	x	x	x	x	x	x	x	x	x	x	x	x	x
Los Angeles	x	x	x	x	x	x	x	x	o	o	x	x	o	x	x	x
Luxembourg	o	o	o	o	o	o	o	o	o	o	o	o	o	x	o	o
Lyons	o	x	o	o	o	o	o	x	o	o	o	o	o	o	o	o
Madrid	x	x	x	o	o	o	o	o	o	o	o	x	x	x	x	x
Manila	x	o	o	o	o	o	x	o	o	o	o	x	o	o	o	x
Melbourne	o	o	o	o	o	o	x	o	o	o	o	o	o	o	x	x
Mexico City	x	x	o	o	x	x	o	o	o	o	o	x	o	x	o	x
Miami	x	x	o	x	x	o	o	x	o	o	x	x	o	o	o	x
Milan	x	x	x	o	o	x	x	o	o	o	o	x	x	x	x	x
Minneapolis	o	o	o	o	o	o	o	o	o	o	o	o	o	o	x	x
Montreal	o	x	x	o	o	o	x	x	o	o	o	o	o	x	x	x

Table 2.2 *continued*

City	Source: F1	F2	PK	DK	SS	FG	NO	RP	HY	CO	TH	KA	LP	RE	SK	GW
Moscow	o	o	o	x	o	x	o	o	x	o	o	o	o	o	o	x
Mumbai	o	o	o	x	o	o	o	o	o	o	o	x	o	x	o	o
Munich	o	x	x	o	o	o	o	x	o	o	o	o	o	o	x	x
Nagoya	o	o	x	o	o	o	o	o	o	o	o	o	o	o	o	o
New York	x	x	x	x	x	x	x	x	x	x	x	x	x	x	x	x
Osaka	o	x	x	o	o	o	o	x	o	x	o	x	o	x	x	x
Panama City	o	o	o	o	o	o	o	o	o	o	o	o	o	x	o	o
Paris	x	x	x	x	x	x	x	x	x	x	x	x	x	x	x	x
Philadelphia	o	o	x	o	o	o	x	o	o	o	o	o	o	o	o	o
Portland	o	o	o	o	o	o	x	o	o	o	o	o	o	o	o	o
Prague	o	o	o	o	o	o	o	o	o	o	o	o	o	o	o	x
Rio de Janeiro	x	o	o	x	o	o	o	o	o	o	o	x	o	x	o	o
Rome	o	o	x	o	o	o	o	o	o	o	o	o	x	x	x	x
Rotterdam	x	o	o	o	o	o	o	x	o	o	o	x	o	o	o	o
San Francisco	x	x	x	o	o	x	x	o	o	o	x	x	o	x	x	x
Santiago	o	o	o	o	o	o	x	o	o	o	o	o	o	o	o	x
São Paulo	x	x	o	x	x	x	x	x	o	o	o	x	o	x	x	x
Seattle	o	x	o	o	o	o	x	o	o	o	o	o	o	o	o	o
Seoul	x	x	o	o	o	o	x	o	o	o	o	x	o	x	x	x
Shanghai	o	o	o	o	o	o	o	x	o	o	o	o	o	o	o	x
Singapore	x	x	o	x	o	x	x	x	o	o	x	x	o	x	o	x
Stockholm	o	o	x	o	o	x	x	o	o	o	o	o	o	o	o	x
Stuttgart	o	o	o	o	o	o	o	x	o	o	o	o	o	o	x	o
Sydney	x	x	o	x	x	x	x	x	o	o	x	x	o	x	x	x
Taipei	x	o	o	o	o	o	o	o	o	o	o	x	o	x	o	x
Tel Aviv	o	o	o	o	o	x	o	o	o	o	o	o	o	o	o	o
Tokyo	x	x	x	x	x	x	x	x	x	x	x	x	x	x	x	x
Toronto	x	x	x	o	x	x	x	x	o	o	o	x	o	x	x	x
Vancouver	o	x	o	o	o	o	x	x	o	o	o	o	o	o	o	o
Vienna	x	o	o	o	o	o	x	o	o	o	o	x	o	x	o	o
Warsaw	o	o	o	o	o	o	o	o	o	o	o	o	o	o	o	x
Washington, DC	o	o	x	o	o	o	o	o	o	o	o	o	o	o	o	x
Zurich	x	x	x	x	x	x	x	x	o	x	o	x	x	x	x	x

Sources:
F1: Friedmann (1986: table 1)
F2: Friedmann (1995: table 2.1)
PK: Knox (1995: figure 1.1)
DK: Keeling (1995: table 7.1)
SS: Sassen (1994, from chapters 1 and 2)
FG: Finnie (1998), adapted in Graham (1999, Figure 1)
NO: Nomura (in Rimmer, 1991: figure 4.1)
RP: Petrella (1995: 21)
HY: Hymer (1972: 50)
CO: Cohen (1981: 308)
TH: Thrift (1999: 70)
KA: Knox and Agnew (1989: figure 2.18)
LP: London Planning Advisory Council (1991: figure 1.2)
RE: Reed (1981: 59–60)
SK: Short and Kim (1999: table 3.10)
GW: Beaverstock *et al.* (1999a: table 2)

the two cities in the business operation. On the other hand, the Manchester people may consider that to expedite matters they need a level of legal expertise that can be found only in London law firms. Thus, all inter-jurisdictional contract negotiations between Manchester and Brisbane will go via London. The point is that with the enabling technologies of IT, old hierarchies can be bypassed but they can also be reinforced. In this new global space of flows there may be hierarchical tendencies within *some* inter-city relations, but for others there will be *direct* links between cities irrespective of their relative rankings. The result is likely to be very intricate patterns when studied across many world cities. In other words, this argument follows Pred's (1977, 1978) identification of spatial organizational complexity in 'national' urban systems, which was described in the previous chapter, and multiplies this complexity manyfold. My favourite portrayal of this is Nigel Thrift's (1999: 272) 'blizzard of transactions' creating a contemporary world-economy of 'unimaginable complexity' (ibid.: 274).

There is a key implication that derives from this argument: it is quite problematic to identify a small subset of 'global cities' if this is interpreted as meaning cities in which global processes are immanent. This idea has led to affiliated concepts of 'global cities' such as 'subglobal cities' or even 'non-global cities'. The point is that under conditions of contemporary globalization all cities are globalizing (Marcuse and van Kempen, 2000). The global space of flows is not an exclusive preserve of large metropolises (Warf and Erickson 1996). They may have locational advantages in terms of enhanced infrastructural environments, and these are important, but they are matters of degree. The experience of cyberspace is not essentially hierarchical; it operates as innumerable networks, albeit across an uneven globalization. In this sense, therefore, all cities are global: they operate in a contemporary space of flows that enables them to have a global reach when circumstances require such connections. It may be convenient to keep the concept of 'global city' for those leading metropolises that have very many global connections as part of their everyday operations, but any idea of 'non-global' cities needs to be banished. Here I use the phrase 'cities in globalization' to convey the idea that contemporary globalization impinges within the operations of all cities and that all cities contribute to the ongoing processes that are globalization.

Jacobs' dynamic cities

In economic life there are, according to Jane Jacobs (1984), two types of urban settlement: those that are dynamic and those that are relatively static. The former expand economic life, the latter are merely spectators at the feast.

To be a spectator is to be an entrepôt, a passive locale through which goods are conveyed but with little associated economic activity. For instance, in the colonial era major port cities evolved, at least one in every colony, as trans-shipment points for transferring the colony's production to the metropole. These cities were simply part of the imperial infrastructure for extracting economic value from a colony. In post-colonial times many became independent capital cities but they remained 'economic sinks' through which the country's wealth was sucked from the newly independent nation. These Third World primate cities are the hubs of the process famously described by Gunder Frank (1969: 6–7) as the development of underdevelopment. Within a framework of economic dependency, these cities are static in the sense that they make no contribution to increase in overall economic wealth. They are involved in a two-way trading system – importing producer goods (e.g. mining equipment, fertilizer), exporting raw materials (metals, cash crops) – but their trade remains based on the production of others. In world-systems terms these cities are peripheral places. The capitalist world-economy has expanded geographically through such periphery formation but economic expansion has required core formation based upon dynamic cities that use new peripheral production for their own economic gain.

Import replacement by cities

For Jacobs (1984), dynamic cities expand economic life. They are centres of innovation in which production feeds directly into trade. In particular, the means of expansion is import replacement. A key process is therefore the conversion of an entrepôt into an innovative centre. Boston, for instance, was, for more than a century, primarily a passive place through which north-east American trade with the British metropole was conducted. The trans-shipments consisted of raw materials such as furs out and consumer goods such as tea in, with few or no Boston commodities leaving town. This suited the colonial masters, of course, who wished this system to continue indefinitely. The political change to independence did not immediately alter the dependency condition, but gradually Boston changed. Instead of simply buying producer and consumer goods from across the Atlantic, it began to manufacture some itself. Replacing European imports did not just mean reducing the city's dependence; now Boston commodities could be traded to other parts of the new United States. This is the route from dependence to an active and therefore real (that is, economic) independence.

Import replacement is a city process because it requires special locales to be generated and sustained. Only cities can embody a critical mass of people and ideas with the skills and flexibility to create the necessary new production. Such production does not have to be only of finished goods; the various producer goods and services that are required can be part of the vibrant new economy. Such a circumstance cannot be simply planned: that is, simply put in place as a 'collection of things' deemed necessary for development. Jacobs uses the economic policies of Peter the Great and the Shah of Iran as classic cases of this misapprehension. Building St Petersburg as a copy of successful Western European cities has been described as the first development project of the modern era (Berman 1988). But St Petersburg never became the 'new Amsterdam' as intended; rather, without the import replacement processes, the city became an 'economic parasite' living off the rest of Russia's economy as the country became even poorer compared to the West. Jacobs goes into some detail on the Shah of Iran's later (1970s) attempts similarly to modernize by buying in Western technology. The Shah's abrupt political end is closely associated with the failure of his economic reforms. Quite simply, you cannot plan, then buy off the shelf, the necessary attributes of a dynamic city.

It is not just the complexity of the internal workings of cities that makes planning for 'modernity' impossible (Taylor 1999: 16–18). Import replacement is self-evidently not a matter for one city alone; it requires connections with other cities. Returning to the example above, in the first half of the nineteenth century Boston became a vibrant, dynamic city, but not on its own. For import replacement to create a net gain in economic wealth there has to be a network of cities. If Boston had merely made for itself the consumer goods it replaced from Britain there would have been no overall gain – some jobs lost in Britain and gained in Boston. The key point is that import replacement is not a condition, it is a process. As Boston's economy became more sophisticated, it needed to buy further producer goods from Britain, boosting production there, while its exports to other US cities became victims of new import substitution production in, say, Cincinnati. Boston's gain is a wider US gain and also Britain's gain. Similar processes can be seen in the rise of Pacific Asian cities in the late twentieth century, with, for instance, Hong Kong replacing economic dependence by relations of interdependence in the world-economy.

Contrasting eighteenth-century St Petersburg with nineteenth-century Boston, and more recent Iranian cities with Pacific Asian cities, Jacobs (1984: 140) concludes that development is a 'do-it-yourself process' that constitutes city network formation. There are the two elements to the process: improvisation to replace imports, and inter-city trading for mutual benefits. The historical record shows that all development successes incorporate these two elements. Improvisation within cities has created a European history of city network formations as a series of successive wealth zones. Jacobs begins her story in post-Roman Western Europe, in which there is a dearth of dynamic cities. However, import replacement in Venice of goods from Constantinople, the 'new Rome' that survived the demise of the western Roman Empire, converted Venice's economic position from eastern dependence to an interdependence involving other Italian cities. It is this first 'modern' European city network that gradually spread its trading to other parts of Europe, eventually leading to import replacement in the Low Countries (Bruges) and beyond (the German Hanse), as described in the previous chapter. And, of course, in the north it is Braudel's (1984) world-cities, first Antwerp and then Amsterdam, that took on the Venice role in stimulating a mutually reinforcing city network that has survived through to the present. Subsequent import replacement and consequent city network formation occurred in Britain, central Europe, the United States and Japan, with Tokyo being a later 'new Venice'.

Cores and cycles

What Jacobs is describing here is nothing less than Wallerstein's (1979) world-systems core formation. He describes core zones of the world-economy as regions that have a predominance of core processes, economic activities that involve relatively high-tech and high-wage production. These zones have been agro-industrial cores in city-rich regions of the world. They are rich in cities because the innovations and organizations necessary for core zones to develop operate through cities. Thus core formation is generated within cities in networks, as Jacobs describes. There can be no cores without city networks; they are the necessary prerequisite of the creation of core zones in the world-economy. In Wallerstein's work cities are neglected and, in fact, many people who use his core–periphery model have delineated his zones as combinations of states. However, this is not what Wallerstein had in mind in defining economic concepts such as core and periphery. For instance, he explicitly divides early modern France into two zones so that only northern France is allocated to the core zone of the period. Subsequently it can be noted that rural Ireland in the nineteenth century, and the ex-Confederate South for much of the twentieth century, both existed as part of the most important state of their times but both were also obviously non-core zones. In other words, they were regions without dynamic cities. Clearly, the core zone must be defined by its dynamic cities, not by important states. The latest expression of this process of city-led core formation has been the rise of Pacific Asia through its cities. Quite simply, the geohistory of city network formation is the geohistory of economic core formation in the world-economy.

Development is not eternal. Nothing stands still in a process; a city that stops innovating will find itself outside the process and therefore returned to the category of static settlement. Many of the industrial metropolises of the late nineteenth century in the north-eastern United States and north-western Europe, great import replacers in their time, found themselves in just such a predicament in the late twentieth century. Such fluctuations in the fortunes of cities are inevitable, but Jacobs (1984) argues that such problems may be eventually overcome if cities within a network are at different positions within economic cycles. Growth via import replacement will continue in part of the network, providing opportunities for other cities to improvise and get back into the process. Jacobs uses a sporting metaphor to illustrate the point: in any team, players need each other simply to compete in the game, but beyond that, at any one time, players having a bad game need to be compensated by other players having a particularly good game to keep the team competitive (ibid.: 209). A key requirement for a city network is, therefore, for it to work in a 'nonsynchronized fashion' (ibid.: 210), so some cities are freeloading on the economic growth as others are delivering it. It follows that the demise of a city network would come about when cities are moving together in an economic cycle so that the whole network simultaneously reaches a static phase. Jacobs mentions deep international depressions as possible contexts for this doomsday scenario (ibid.: 211). This is Jacobs' economic nightmare:

> If global [meaning worldwide] city stagnation ever does occur, it will inexorably cause economic life to stagnate and deteriorate, and there will be no way out: no existing vigorous cities to intervene, no young cities arising while they still have the opportunity to do so.
>
> (ibid.: 134)

It might be conjectured that contemporary globalization and its world city network provide the most potent recipe for a possible synchronized economic disaster since the beginnings of the modern world-system. I develop such speculation in the final chapter.

A pure space economy

Jacobs' nightmare is nothing less than a peripheralization of the core. This is a reminder that there is always much more to the world-economy than the regions of the core, even though scholarship is usually disproportionately focused on the latter. I need to return to the 'static settlements' and see how they fit into the world-economy. Import replacement and city network formation define the core, but what of the rest of world-economy, the regions without dynamic cities? In this section I describe Jacobs' ideas on the latter and combine them with the previous argument to create a model of the space economy. I call it 'pure' because, for the moment, I wish to create a world model of economic processes with no political inputs. This is obviously an extremely unreal space economy, one without boundary divisions, but it will help us to understand the role of states when I come to add political processes to the configuration in the next section.

City and hinterland

Starting with the core, Jacobs (1984: 41) argues that 'any settlement that becomes good at import-replacing *becomes* a city' (emphasis in the original). For this to occur at a significant level, there are five 'great economic forces of expansion' (ibid.: 47) that are unleashed on the economic landscape in which the city exists. These forms of growth are as follows:

1 *new market force*: a rapid increase in the size of the city market for new imports through the city network;
2 *new employment force*: a rapid increase in the quantity and diversity of city jobs associated with import replacement;
3 *relocation force*: a more rapid turnover in city industries as older enterprises move out of the city;
4 *new technology force*: an increased use of new city technologies; and
5 *new capital force*: a generation of new city capital.

These five forces operate to make the city dynamic, but their influence is not restricted to the city.

In the hinterland of an import-replacing city the five forces of growth act in chorus to generate a dynamic city region dependent on the central city for its vibrancy. The increases in market, jobs, technology, transplanted industries and capital create a mix of opportunities that lead to the diffusion of the city's economy. For this to happen, the economic forces have to complement each other so that their influences are mutually reinforcing. For example, when technology improves agricultural production, causing rural unemployment, there is new job creation in the city to absorb the workers, pushing ex-farmworkers into more productive employment. At the time of her writing, Jacobs identified the Tokyo region as especially powerful in the way it was shaping and expanding a new city region. Today she might use as examples Greater New York spilling over into New Jersey and Connecticut, London engulfing all of south-east England, Hong Kong reaching out into the Chinese mainland, and the new region around San Francisco Bay. Not all cities develop such city regions, but Jacobs argues that 'the very mechanics of city import replacing automatically decree the formation of city regions' (1984: 47). This suggests that major cities lacking city regions – she mentions Glasgow, Cardiff, Madrid, Naples and Montevideo as examples – have not fully incorporated the dynamics of import replacement into their city economies. No matter; for now it is enough to appreciate that dotted across the world there are spaces that can be identified as city regions that coalesce in broad zones to create the contemporary core of the world-economy.

Grotesque economies

The projection of the influence of economically dynamic cities is not limited to the city hinterlands. The five processes can be observed 'rippling out' (Jacobs 1984: 43) across the space economy far beyond the city regions. However, whereas the city regions experience the full panoply of the five forces, beyond these spaces this nexus of growth processes unravels. Each force produces its own specific form of economic change, resulting in five types of 'unbalanced' spaces in the economic landscape. Thus do 'cities shape stunted and bizarre economies in distant regions' (ibid.: 59). These 'economic grotesques' (ibid.), as Jacobs aptly terms them, make up the bulk of the periphery of the world-economy. The five types of peripheral spaces are supply regions created by city markets, abandoned regions created by city jobs, cleared regions created by city technologies, transplant regions created by exported city industries, and cityless regions distorted by city capital. I am going to argue that these five types of space actually divide into two groups: three types of disrupted places (abandoned regions, cleared regions and warped cityless regions), and two types of connected regions (supply regions and transplant regions).

The disrupted places are each typified by an extreme perverse destruction of their economic landscape. First, in the case of *abandoned regions* the attraction of city jobs is depopulating regions as young people leave to find work in vibrant city regions. Jacobs gives the examples of Mexico, southern Italy, Yugoslavia and Turkey, where economic remittances from emigrants had become important sources of wealth. Today the cities of the European Union, the United States, Canada and Australia are acting as magnets for economic migrants from many parts of Eastern Europe, North Africa and the Middle East, Latin America, and East Asia. The economies left behind are thus shrinking but otherwise remain as stagnant regions with static cities. Typically they will be recipients of cash remittances from their exiled sons and daughters that may bring pockets of relative wealth; but wealth transfer is not wealth creation. Such a process requires dynamic cities and it is because these are not to be found in the region that the sons and daughters left the region in the first place.

Second, *cleared regions* are where technology has destroyed the jobs in a region so that all but a few workers are forced to emigrate. The Scottish Highland clearances of the eighteenth and nineteenth centuries are the classic case, where new husbandry involving improved livestock led to the demise of the crofting tradition. Jacobs (1984) also gives the example of the depopulation of the rural US South in the first half of the twentieth century. In a region with no cities generating new jobs, the rural exodus had to be a regional exodus: to northern industrial cities. Today, anti-peasant new technologies have been a cause of the huge rural migrations producing mega-cities in the semi-periphery and periphery zones of the world-economy. As Castells (1977) pointed out many years ago, Third World urbanization in the second half of the twentieth century has broken the link between urbanization and industrialization that first formed in industrial core regions in the nineteenth and early twentieth centuries. In other words, in the core formation process there were new jobs waiting for the rural migrants; not so in what have become the great mega-cities of today's periphery. The key point, therefore, is that clearances are not linked to new job-creating cities.

Third, *cityless regions* are severely disrupted when city capital is used to invest in rural regions that do not have the basic infrastructures – no city regions – to absorb and use the capital productively. Jacobs (1984) uses as an example the proliferation of dams in so-called development projects that destroy peasant farms by flooding but never repay with the economic development that is promised. The Volta Dam in Ghana is the classic example here; it produces masses of electricity in a region unable to consume it. A transplanted aluminium plant apart, there is nothing to take advantage of the cheap electricity. Jacobs also shows how the much-heralded Tennessee Valley Authority (TVA) is just as much a failure: Knoxville, the city housing the TVA headquarters, has not been able to harness any import replacement processes and the region has degenerated into an impoverished area operating as a simple supply region for electricity. This mania for dams and development continues to the present day, of course. Jacobs sums up the stupidity as follows: if modern cities, and thus modern civilization, die, future archaeologists will find dams to be modernity's most impressive memorials (ibid.: 122)!

The connected regions are two well-known categories. *Supply regions* are typically specialized in just one or two commodities that make them very vulnerable to changes in city markets. The economic well-being of Ghana's people, for instance, has fluctuated in recent years with the price of cocoa futures on Chicago's commodity markets. The really classic city example that Jacobs convincingly argues is the impoverishment of Montevideo. In the mid-twentieth century the small Latin American country of Uruguay was relatively rich, its wealth being derived from meat products exported to cities in Northern Europe and the United States. The combination of its size and wealth led to claims that it was the 'Switzerland of Latin America' (Jacobs 1984: 60). However, the two countries could hardly be more different in terms of their cities: Montevideo never developed any of the dynamic import-replacing forces that are found in Swiss cities. In the good times it had no need to: it operated as a great entrepôt for animal products, and that alone temporarily made it a rich city. However, dependent on distant city markets, Montevideo's affluence soon waned.

In a very similar manner, *transplant regions* are highly dependent on distant cities. Jacobs (1984) uses the example of San Juan in Puerto Rico as the example where transplanted industries, attracted by low labour costs, have failed to generate an economically vibrant city. I have illustrated this before: transplanting city industry is, of course, what Peter the Great and the Shah of Iran attempted in their failed modernization schemes. As I argue in the case of those historical examples, transplanting is simply not enough.

Finally, there are peripheral spaces that are at the opposite end of the world-economy's wealth creation spectrum: *bypassed places*, regions that have lost their cities and city ties or have never historically developed them. This is the world of subsistence economies, the rural backwaters of the Third World. For Jacobs (1984), the classic example is Ethiopia, whose cities stagnated some two thousand years ago. Jacobs comments that the stagnation is not exceptional, but 'what is remarkable, rather, is that once this economy had lost its cities and city ties, it has remained by-passed for so long' (p. 130). This case seems to be a prime source for her economic nightmare related in the previous section: the whole world becoming 'Ethiopianized' – that is, bypassed into a cityless future.

Core, periphery and semi-periphery

These various spaces can be put together like a jigsaw to construct the space economy of the contemporary world-economy. There is no published city-centric economic world geography but I can broadly depict the pattern as follows. At the core of the world economy is a network of world cities with their city regions. As Jacobs (1984) describes them, the rest of the economic spaces form different sections of the periphery. However, this formula removes the dynamism within the world-geography, the possibility of regions rising and falling. This is certainly not her intended outcome. In fact, Jacobs counters the pessimism about most of the world engendered by her analysis, pointing out that it is possible for new import-replacing cities to arise in such peripheral locales (ibid.: 133). Of course, they can arise only through their strong connections with the core's city network. Thus it is not surprising that she only identifies examples for the connected group of spaces: supply regions from which Hong Kong, Seoul and Singapore have recently grown, and transplant regions out of one of which Taipei has recently emerged. I think it is better to conceptualize these two types of spaces as both being supply regions, i.e. primary supply regions and secondary supply regions respectively. The fact that also all Jacobs' historical examples of the emergence of new import replacement cities – from Venice to Tokyo – are from supply regions suggests that such regions have a potential for growth not found in the other types of peripheral spaces.

In Wallerstein's (1979) schema the dynamic role – the world zone where regions rise and fall – is taken by the semi-periphery category. These are places wherein core-producing and periphery-producing processes are balanced. That is to say, there are some relatively high-tech, high-wage economic activities occurring alongside more traditional low-tech, low-wage activities. In terms of Jacobs' (1984) types of spaces, it is the supply regions through their entrepôts that have the potential for rising through to core status. This means changing from a 'dead end supply region' (p. 143) – that is, a region not *in the network* but *attached to it as a few simple links* – to becoming an integral *part of the network* connected *through multifarious linkages*. This is what the dynamic Pacific Asian cities mentioned above all achieved in the second half of the twentieth century. In this period Pacific Asia as semi-periphery encompassed a collection of different supply regions, some dead ends (peripheral processes) and others developing import-replacing cities (core-producing processes). This is how the semi-periphery is defined through its cities, which today means world cities.

Before I leave this discussion I must reinforce the city-centred nature of the approach by distancing the argument from the popular 'import substitution' development policies of many poorer countries in the past few decades. What is the difference between import replacement as defined here and import substitution in development studies? The key point is not only that the former operates through states and the latter through cities, but that import replacement is simply not a policy, but an economic process. That process requires cities. And it makes no difference whether a city is replacing commodities from foreign or domestic cities (Jacobs 1984: 43). The process remains an economic one, not a matter of political geography. In contrast, a policy of import substitution dissipated across a territory controlled by a state is a planning instrument destined to go the way of all state planning that engages an ever-changing modernity. However, states do more than devise failed development policies; it is time they were brought fully into the argument.

A political economy world

The pure space economy developed above is a very simple abstraction, a world of economic processes without political processes. The modern world is actually divided into nation-states through which most politics is channelled. Thus the reality of the space economy is that it is criss-crossed by political boundaries. Using Castells' (1996) language, there is an economic space of flows, notably city networks, coexisting with a political space of places, notably the mosaic of states. This is a political economy world.

This section rehearses two important arguments by Jacobs (1984) for bringing the state back into the discussion. First, I address the idea that the state is the prime entity through which economic processes operate. Here I conclude that the concept of a national economy is a myth. Second, I note that whereas there may be no such thing as a national economy, there are, nevertheless, state apparatuses whose prime functions are economic. Thus states are important as economic players and they can have profound effects on the cities within their boundaries.

The myth of the national economy

The fundamental division between Jacobs (1984) and conventional economics reduces to a geographical question: what is the basic entity through which economic processes operate? The orthodox answer – whether the orthodoxy is traditional political economy or Marxist economics, Keynesianism or monetarism – is that the state defines the economy. For Jacobs (1984: 31), this is simply an odd answer: states are 'political and military entities' and 'it doesn't necessarily follow from this that they are also the basic, salient entities of economic life'. In fact, it is highly improbable that the vicissitudes of politics, producing as it does a quite haphazard mosaic of states ranging in size from China and India to very many small states such as Luxembourg and Lesotho, should delimit 'economies'. Hence despite the commonplace references to the 'national economy' in both the media and academia, for Jacobs this is a gigantic myth distorting economic thinking. She traces it back four hundred years to simplistic mercantilist theory that promoted the hoarding of gold within a country. This 'mercantilist tautology' – studying only economies in states and thereby assuming that states define economies – was taken on board by the 'father of economics' (and mercantilist critic) Adam Smith in his *Wealth of Nations*, and the myth has been transferred across economic schools of thought through to contemporary econometrics.

What are these economic spaces that go by the name of 'national economy'? Jacobs (1984: 32) is scathing about their veracity: she dismisses them as 'grab bags of very different economies, rich regions and poor ones within the same nation'. Unlike city economies, 'national economies' are not wealth creators; they do not have the ability to shape the space economy by projecting economic changes, both growth and stagnation, across long distances. Only cities can do this (ibid.), which is why Jacobs focuses on them in creating her pure space economy. 'National economies' are divisions of the actual space economy that combine city regions with other types of economic spaces. Thus all but the smallest states have their 'regional problem'. The most famous is the 'two Italys', consisting of dynamic northern cities and southern static cities and abandoned regions. Previously I used equally classic cases: the abandoned regions of Ireland within the UK economic space (a century and a half ago) and the abandoned regions of the American South in the US economic space (half a century ago). This variety of types of spaces is important not just because it illustrates the economic arbitrariness of 'national economies', but particularly because it creates a politics wherein economic policy cannot properly respond to economic indicators of change.

All economic policies are like big experiments in systems control. The policy makers expect that a given input into the economy will produce a desired output. Development models, for example, provided a series of necessary steps to take in order to produce a 'developed nation'. The problem is that there has never been an example of such a policy producing this desired output. This is because the 'nation' expecting to become developed was not itself an economy, so that input measures and output measures are both aggregations based upon an amalgam of information covering different sorts of economic spaces. Not surprisingly, policies typically have different effects in different parts of the country, so that in testing the success of a policy, the aggregate output indicator for the 'national economy' is a grossly flawed number. Quite simply, for all 'national economies', whether deemed 'developed' or 'developing', there are serious feedback flaws in economic policy. Jacobs (1984) highlights the problem in the richer countries of the world that were experiencing confusing feedback at the time of her writing. This was 'stagflation': inflation and unemployment rising together to nullify Keynesian policy levers that used increased inflation as input to produce the desired output of reduced unemployment. Governments at the time simply turned around the formula (raise unemployment to reduce inflation, or monetarist policy), but Jacobs argued that the whole approach was wrong. Cities and city regions are the basic economic entities and it is only with these units that any meaningful feedback between input and output is possible.

Creating primate cities in the core

In Chapter 1 the rank-size rule was derived from systems thinking to indicate the existence of a 'national urban system' within a sophisticated national economy. The primate city model was then interpreted as the counter-case: a national city distribution in which few economic forces were operating, so the leading city grew at the expense of the rest. The latter process was commonly found in Third World countries, suggesting that the two types of city distribution could be combined in a developmental model (Berry 1961). This idea works very well for countries in the Americas: the United States and Canada have rank-size distributions, Latin American countries nearly all have simple primate distributions. But the model does not translate well to Europe, where the United Kingdom and France famously have very strong primate patterns. The disproportionate sizes of London and Paris can be explained by their historic roles as imperial capitals, but this does not account for the frequency of primate patterns among other European countries. Boulding (1978: 151) refers more generally to a 'sharp distinction' between capital cities and provincial cities. Copenhagen and Stockholm, for instance, are primate cities in Denmark and Sweden respectively but neither was ever a great imperial centre. By drawing on Jacobs (1984) it can be shown that rather than being anomalies, these European primate city distributions are the norm, leaving the rank-size rule as the counter-case.

The richer countries of the world by definition consist primarily of amalgams of city economies. Although such 'national economies' are a myth, the fact that these economic spaces are nevertheless subject to economic policies cannot be ignored. However, it should not be thought that such 'national' policies are automatically neutral, or random, in their influence across the city economies within a state. In a situation where one city begins as the largest city-economy, usually the capital city, 'national economy' feedback will more closely reflect this city economy's workings that any other. Hence, what feedback is obtained will suggest policies that broadly incorporate the economic needs of this one city at the expense of the other cities. Over time, therefore, economic policies will tend to favour that city simply because its economic circumstances are more and more like the aggregate 'national economic' statistics. Important national policy, for instance in interest rates to control currency levels, will respond to the requirements of a state's largest city-economy at the expense of the rest. The geographical outcome of this political economy process is to create a situation where one city increasingly dominates: the aggregate 'national economy' gradually evolves towards a large city economy, with other city economies reduced to being 'static' territorial appendages.

This is precisely the urban geography that is found in most European countries. Typical 'national economies' that mimic a 'one-city economy' are Sweden/Stockholm, Denmark/Copenhagen, Norway/Oslo, Finland/Helsinki, Austria/Vienna, Greece/Athens, Portugal/Lisbon and, of course, France/Paris and the United Kingdom/London. For instance, in the last-named case, successive governments have pursued policies favouring the rich South-East – London's city region – at the expense of what remains of the great Victorian cities of northern Britain. In 1999, for instance, the Governor of the Bank of England, charged with making British monetary policy, declared that it was worthwhile increasing unemployment in north-east England, Newcastle upon Tyne's region, if it reduced inflation in the South-East, London's region. This is a classic statement of what Jacobs (1984: 203) calls a 'city-killing'. However, this process is less likely to occur where there is a more decentralized state (e.g. Switzerland), and where the state is large enough to accommodate numerous large city economies such as in the United States and Germany. This appears to be the explanation for the rare Third World primate city distributions, such as in India and China, that are sometimes viewed as 'anomalies' because they are more 'rank-size' than primate. In all large countries, political economy policies will impact more neutrally on the various city economies over time, so no one city will necessarily grow to dominate and create a primate pattern. In other words, there will be no economic policy customized to the needs of any one city economy. This is important for cities that have to restructure their economies (not simply 'reinvent' themselves via a city boosterism) and need the helping hand of an economic policy; for example, without its own policies, the Pittsburgh city economy may be just as badly off as the Manchester city economy.

Despite this severe disadvantage for some cities, for the favoured primates state policies do have a strong positive effect. But not necessarily for the long term: the primate-making tendency runs counter to the advantages of a non-synchronized city network described previously. With primate city growth fuelled by its state, the cyclical replenishing of growth processes is destroyed, at least within the state's territory. A long-term continuation of primate city growth within nationally governed economic spaces (aka 'national economies') would likely finish up with a world mosaic of territories each containing only one dynamic city. This is a route to Jacobs' no cities nightmare. I return to this future scenario in the final chapter.

World cities as the new dynamic cities

I have gone back to basics to learn to think relationally about cities in new ways. This was necessary for two allied reasons. First, the only major sustained research effort at understanding inter-city relations, the national urban systems school, had placed hierarchy at the centre of thinking about how cities relate to each other. However, this rather rigid view of cities did not seem to me to be how cities were necessarily connected at the global scale, given the enabling technologies that had helped create contemporary globalization. Second, because of the dearth of evidence on connections between world cities, hierarchy had come to dominate thinking on these new inter-city relations as well. 'World cities hierarchy by default' was not a viable starting position for this book. And so I needed some fundamental ideas from outside the national urban systems literature to provide new insights for the world cities literature. I found this in the seminal work of Jane Jacobs. The question for this brief conclusion is, what has been learned?

I take forward the following ideas. First and foremost, I have been able to use Jacobs' ideas to see the limitations of transferring 'national hierarchy' to 'global hierarchy'. By and large, cities do not operate by command: inter-city relations are not primarily about cities directly controlling, or being controlled by, other cities. The exceptions are political cities – state capitals and lower-level administrative centres – whose functions mirror a bureaucratic framework of strict hierarchies of authority and power. If cities were only the creatures of states, then studying them as national urban hierarchies would be sensible. But globalization has reasserted that cities are more than subunits of states, more than even the 'powerhouses' of 'national economies'; they are their own economic entity within transnational spaces of flows. Cities operate through networks, patterns of inter-city relations in which trading cities complement each other within spaces of flows. Jacobs shows that it is through such networks that economic growth has spread historically and it is the contention of this book that, within contemporary globalization, there is now a world city network.

World cities are therefore actual, recent or potential 'dynamic cities', as defined by Jacobs (1984), depending on their cyclical location in the network formation. They are centres for new productions that increase overall economic wealth within the world-economy. Thus rather than looking to headquarters economic functions of world cities – as command and control centres of large corporations – it is, according to Jacobs (1984), the innovations generated within cities that should be the first focus of attention. This idea is indeed central to the world city literature, in which these cities provide a stimulating milieu, variously conceptualized, for information, knowledge and creativity to intersect in the production of new service commodities. This 'internal' nature of world cities is not my prime subject here, but Jacob's second required focus for dynamic cities is how these new commodities circulate between the cities to create a network. Of course, these two topics cannot be separated in reality and therefore I use existing ideas on the first focus as the basis for my contribution to kick-starting research on the second focus.

The production of world cities as Jacobsean dynamic cities is to be found in the service sector. I use Sassen's identification of advanced producer services as a key production process of her global cities as my starting point. The reason for this choice is because these firms offering services have taken advantage of IT to create world cities as strategic places for servicing global capital. Some of these services are old, such as merchant banking and maritime insurance; others are more recent, such as commercial law and company auditing, while still others are quite recent such as advertising and management consultancy, but these trading city functions all have in common a propensity to have globalized their services across many cities. Such global practice is a crucial part of the innovation at the heart of the new global services. Whereas in earlier networked cities high prices derived from geographical monopolies (e.g. in the spice trade), today high prices derive from professional knowledge monopolies (e.g. inter-jurisdictional law). It is such knowledge that is being traded through the world cities to create contemporary dynamic cities and their world network. This is my starting point for considering how the world city network can be conceptualized, measured and analysed.

 **Part II** **Connections**

3 ▶ Networks of cities

> An image thus begins to form of the city as a 'node' of global networks, where local identity and the urban territory, as a stratified deposit of natural and cultural assets, no longer have value for what they are but for what they become in the processes of valorisation. The partial truth contained in this image is that the city as local society is no longer identifiable for its stable embeddedness in a given territorial mileau. It is instead a changing connective configuration with variable actors which can be thought of as 'nodes' of local and global networks.
>
> (Dematteis 2000: 63)

Going 'back to basics' translates into searching out city networks. This leads to many new questions, not least of which is 'what sort of network?'. In this chapter I provide an answer to this and other questions such as 'how are cities connected to each other?' and 'how can a city's connectivity be measured?'. Asking such questions is itself an important step because past concepts used to describe how world cities relate to one another have been notoriously vague in such matters. As well as Friedmann's familiar 'world city hierarchy', there is reference to, among others, 'global network of cities' (King 1990: 12), 'transnational urban system' (Sassen 1994: 47), 'new global urban hierachy' (Wu 1996: 121), 'functional world city system' (Lo and Yeung 1998: 10) and 'global urban network' (Short and Kim 1999: 38). In using these concepts, none of the authors specifies their construct in a manner that allows answers to be given to the sort of questions just posed. The purpose of this chapter is to present a precise specification of a world city network in order to begin the task of measuring world city network relations.

Specification requires identification of the agencies that create the world city network. In the first section below, key institutions are discussed in the way they interact to generate contemporary cities and their networks. From this argument, one particular agent is chosen as the key instrument for the specification. The vital implication for the second section is that I have to specify an unusual type of network, one where the nodes – cities – are *not* the prime agents of the network formation. I identify the world city network as an interlocking network in which it is the 'sub-nodal' level where the key agents – service firms – are to be found. Specification of this unusual network model stipulates what information and data are required to describe and measure such a network. This is the subject matter of the third section, in which the production of a large data matrix is described. This matrix is the evidential basis upon which the majority of analyses are made in the following chapters. In this chapter the concluding section introduces some initial elementary measurements from the specification.

World city network formation

Formation is a process. Cities and their networks are being continually re-formed. Within the world city literature the focus has been upon world city formation. Here I go a stage further to develop an understanding of world city *network* formation.

On not reifying cities

In the previous chapter cities were identified as the basic economic entities for creating new wealth and stimulating development. I must be careful, however, on how subsequently to interpret these economic entities. There has been a tendency to reify cities – that is to say, to treat them as actors in situations where they do not have agency. The model that is developed in this chapter is built precisely on not treating cities as the prime agency in world city network formation.

Cities, of course, do not of themselves create economies. Abrams (1978) is particularly useful in affirming this and attacks the use of concepts such as 'generative city' and 'parasitic city' as popularized in traditional development literature. In Jacobs' (1984) theoretical framework, cities are either economically dynamic or static. These concepts are intended to be descriptive, thus the terms should not be extrapolated to mean that, say, dynamic cities are 'generative cities'. Dynamic cities are part of a generative process in the formation of city networks but it is not the cities *per se* that are instrumental in the generation of new inter-city relations. The economic links that tie cities together in networks are forged by specific economic actors: historically, merchants trading their commodities across the network. At the most general level, the agents of economic change are the holders of capital (in its many forms), and it is their decisions that are vital for economic growth or stagnation. But, of course, this does not mean that cities are mere platforms for capital-holders to use. Dynamic cities attract capital because of the capital-expanding opportunities that exist there. Conversely, capital leaves static cities because of the dearth of such opportunities. In these circumstances it is relatively easy to slip into a language that reifies cities. For instance, Abrams (1978: 17) takes Braudel to task for sometimes lapsing into this type of thinking.

But the question of reification is not always that clear-cut. Often a city name is used as shorthand for describing all the processes with their agents that are operating through a city. During the Cold War, the contending parties were sometimes recoded as 'Washington' dealing with 'Moscow'. This need not be a problem when it is clearly understood what the use of the city name implies; what its referents are. In the previous chapter 'St Petersburg' was identified as having a 'parasitic' role in the tsarist Russian economy. In this case, the agency is obvious: the tsarist administrations that built and developed the city to such a degree that government agency has been commonly identified with this pampered capital city creation. Similarly, in another example for the previous chapter, it is not 'London' that is 'killing' Newcastle but rather successive UK governments whose 'national' economic policies favour the former. The last two examples remind us that it is not just economic agencies that create cities.

Historically, this question is encompassed in 'origins of cities' debates in which cultural-religious, political-military and economic-mercantile agencies vie for the 'honour' of inventing cities in different regions. In fact, these agencies are all important in specific contexts: religious places, strategic places and marketplaces can each form the basis for developing a city. In reality these different types of agency operate simultaneously to produce quite complex inter-city patterns. An example will elucidate.

In early medieval Europe the revival of cities commonly involved the spread of Christianity (possession of a cathedral is still the traditional defining feature of a city), the institution of political order (secular authorities awarding royal charters to cities), and the resumption of non-local trade (sometimes involving 'free cities' not dependent on the power of their religious and temporal neighbours). In all cases cities are not developed in isolation; every city has relations with other cities. Note, however, that these different agencies imply different patterns of relations. The two non-economic agencies organize space hierarchically, whereas the economic agency creates networks, as Jacobs (1984) describes. The Church is explicitly hierarchical in a bureaucracy that is translated on the ground as levels of diocese each ruled from a city. The political agency becomes similarly hierarchical with the establishment of a permanent capital city, although it never achieved the level of organization found in the Church until the modern era. Note that both these city-making agencies are organizers of space: their spaces of flows between cities – orders down, taxes up – create a city hierarchy aimed at controlling a tamed space of places. In contrast, in the merchant's world the flows between cities – reciprocal trade – created a city network, a dynamic space of flows, necessarily unplanned, to enable economic growth to occur. Clearly, the latter case coincides with the interpretation of cities from the previous chapter, but for the moment the key point is that it is not cities that are making these patterns of inter-city relations: it is the papal bureaucracy, not 'Rome', the French royal household, not 'Paris', and the families of merchants in northern Italy, not 'Venice', that are critical agencies of city hierarchical and network formations in medieval Europe.

The most important consequence of broaching the question of reification is that it forces a search for the specific agents who make cities and their networks. It is necessary, however, in avoiding reification, to make sure I do not throw the baby out with the bathwater (Bagnasco and Le Gales 2000: 30). Cities do have city governments with budgets to improve their 'place', perhaps to make it more attractive to holders of capital. Such city boosterism policies have become commonplace in the modern world. Hence although I must not reify the city as an entity, I can disentangle the unitary view of a city so that London, for example, becomes a complex of actors in which the important ones may impinge decisively on London's position in the world city network. Obviously I would not want to ignore the role of mayors and other city officials and politicians in a specification of the network. This means that the question of agency needs careful elaboration to identify the key actors.

Agency in world city network formation

The makers of cities remain multifarious to the present day. Many contemporary world cities clearly bear the mark of their makers: cultural agency has created Los Angeles as a world media city; political agency has created Geneva as an international institutional city; and economic agency has created Hong Kong as an international financial centre. Of course, the leading world cities are the result of all three of these agencies creating 'well-rounded' world cities, as initially envisaged by Hall (1966) and strongly implied in Sassen's (1991) concept of global city. London and New York are archetypal in this respect: both are very important media, political and financial centres and much else besides. As has already been made clear, in this study, following Jacobs (1984), I focus on the economic agents and their creation of city networks. Following Sassen (1991), the agents I concentrate upon are advanced producer service providers. Thus I follow the traditional urban approach of researching inter-city relations through treating cities as service centres. I specify a world city network of global service centres. However, I do not wish to suggest that service firms act alone in the creation of the world city network.

To understand the formation of the world city network I will need to consider four key agencies. Starting with business service firms, the focus is upon those that export services beyond the local city market. Unlike in the case of other commodities, trading in services usually requires some direct attendance to the buyer/client, so the process of exporting in this case often means expanding office networks to service in many places. From the 1980s many firms have become worldwide in their provision of services and I call them global service firms. These offices are concentrated in cities, as Sassen (1991) describes, and cities as local networks of institutions, economic and political, are the second agency. For instance, when referring to 'London' in this context, I mean both the formal network of institutions and agencies such as the Office of the Greater London Mayor, the City Corporation, the London Assembly, the London boroughs, the Cross River Partnership, the London Pride Partnership, and the informal networks of city practitioners who make the city work. A third important agency of world city network formation is the multiplicity of supervisory institutions that oversee the practice of individuals and firms within particular service sectors. These organizations and institutions provide regulatory frameworks and professional codes of conduct that govern the practices of firms. For some services these rules are very strict and the regulators operate as gatekeepers (e.g. law); in others they are relatively lax (e.g. advertising). The fourth agency consists of the nation-states. This includes the state apparatus, especially that involving economic policy, and general national culture of conducting business and its relation to national society. For the former, the relative liberalization of national legal and economic policies may be crucial in enabling flows between world cities. The best example is London's 'Big Bang' in 1986, when the British government gave foreign investment banks access to the London Stock Exchange following state-led measures that deregulated the British financial system. Equally, the culture of saving in Japan has been vitally important for the growth of Japanese banks.

I suggest that it is these four agencies – service firms, city governments, service-sector institutions and nation-states – that, taken together, are primarily responsible for shaping the world city network. However, these processes do not operate singly to sustain the network; they operate in conjunction with one another in quite complex ways. One way of viewing this is to look at how pairs of agencies interact. Six pairings can be obtained from a list of four. In this case they can be further divided into two types of relation, two causal nexuses and four identity assignments. The causal nexuses involve two-way connections that are necessary for both agencies to reproduce themselves. The identity assignments are one-way relations in which one agency moulds the character of another agency.

The two nexuses are vital relations. *Nexus I* draws together service firms and cities in a relationship of mutual reciprocity. Cities need a critical mass of firms to produce the necessary knowledge environment that the firms critically need. This joint reinforcing process is at the heart of definitions of world cities, and indeed of cities more generally. The process has interesting implications in terms of interests: for instance, Japanese and American banks with major offices in London have a vested interest in the future success of London as an international financial centre, while equally London has a vested interest in 'its' Japanese and American banks doing well. *Nexus II* is the interactions between business service sectors and the nation-states. The professionalization of knowledge-intensive services has been a national process involving the state in either legitimizing self-regulating (national) professional associations or else regulating directly. In turn, these professionals have manned the state apparatus to enable the state to function. This is the original private–public partnership. Traditionally it has been lawyers who have straddled the private–public boundary, followed by bankers, but today accountants, management consultants and, of course, advertisers with their PR skills are also becoming prominent in government circles. In sum, each nexus represents a binding together of agencies in a conjoint exercise.

The four identity relations involve one agency defining the nature of the other agency. The first such relation, and the most obvious, is that between service firms and their sectors. Firms are identified by their sector because of the customs, knowledge and practices governed by sector institutions, notably professional organizations. In this way law partnerships operate in a different professional context as compared with insurance companies, and advertising agencies in a different context as compared with accountancy firms. The second relation is between cities and states. This relates to arguments in Chapter 1 about modern cities being nationalized. It involves another set of customs, knowledge and practices shared by cities within the same country. Even the most 'global' of cities remain characterized in part by their state location: London's pre-eminence cannot be separated from its historical British imperial role, and both New York and Tokyo's pre-eminence relates to their locations within the two largest 'national economies' in the world. The third identity relation is between city and sector. All cities are different, and an important criterion for differentiating them is their mix of services. A city dominated by financial services – an international financial centre with limited non-financial services – is very different from one dominated by, say, advertising. In Europe this difference can be represented by Frankfurt and Barcelona respectively. A more evenly distributed pattern of services may be thought of as a 'full-service world city': New York, London and Paris have the best claims to this label. The final identity relation is between firms and nation-states. Although there is a debate about the possible existence of a 'nation-less firm', in practice all firms have the mark of their national origin in their workings (Dicken 1998). For instance, a German law culture that is statist and intellectual has led to German law firms being unable to globalize in the way that US and British law firms have; the latter work through 'national law cultures' that accept more readily commercial and entrepreneurial ways of thinking about law.

This model of four key agencies producing and reproducing the world city network lays the groundwork for specification. Although each agency is necessary for the network formation to occur, I am not going to treat each one equally in what follows. As indicated earlier, I will privilege the global service firms in the specification. There are two reasons for this. First, and following Jacobs (1984), it is the firms as economic agents that produce the wealth upon which the network has been built and is sustained. Second, it is the firms through their office networks that have created the overall structure of the network. Since the latter is a focus of much of what follows, it is to firms that I look for specifying and measuring the network. Other agencies are not ignored; they play important roles for interpreting network patterns. Figure 3.1 shows the elements of the model that features in discussions of results later: the firm–city nexus is at the heart of the network formation but I supplement it in what follows through interpretations

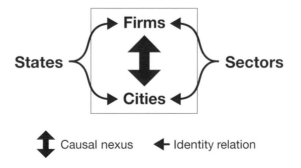

Figure 3.1 Attendants of world city network formation

involving both national and sectoral identities of firms. However, the new global patterns of inter-city relations are ultimately the result of the recent rise of large numbers of global service firms and therefore it is with this agency that I begin.

The world city network as an interlocking network

As shown in Chapter 1, international business was serviced across the world long before the emergence of contemporary globalization. In the mid-twentieth century, for instance, faced by monetary transactions across borders, banks could recommend their clients to use their correspondence banks, favoured banking partners in other countries. Similarly, legal contracts for projects in more than one jurisdiction could be dealt with by law firms having reciprocal relations with firms in other countries. Such practice consisted basically of a series of *ad hoc* arrangements brought into play as and when necessary. Globalization undermined this situation in three fundamental ways.

First, the globalization of clients' affairs meant that to keep an important client, a service provider would have to service where its client wanted to be serviced. Obviously, a major multinational would not want to deal with, say, a different accountancy firm in every country it did business in. Second, as service firms moved from following their clients across the world to developing their own global strategies to win new business, they became concerned to protect their own brand name. The old *ad hoc* arrangements provided no guarantee that clients would be serviced at a level expected in the home country. If things went wrong, this was a recipe for losing valued clients. Providing a direct service allowed for quality control for the service firm and offered a seamless service for the client. Third, the combining of communications with computers created a new IT environment that meant that services could be more easily organized on a multi-office basis. And, of course, for this enabling technology it did not matter whether the expansion office was in a neighbouring city or on the other side of the world. These are the circumstances that spawned global service firms and hence a world city network. I am now in a position to convert discussion of world city network agency into the first step in specification: identification of the type of network.

What sort of network?

Networks are normally specified at two levels: the level of the network itself that defines the scope of the relations, and the nodal level defining the agents whose relations constitute the network (Knoke and Kuklinski 1982). A typical example would be to interpret a street gang at a network level with, at the nodal level, each gang member being an agent of network formation. However, such a two-level model is inappropriate for the world city network because it would cast the cities, as nodes, in the role of prime agents, a position previously rejected. Thus I require a type of network that includes a third 'sub-nodal' level to accommodate the sub-nodal actors – global service firms – that create the network. Hence specification of this network is not so straightforward. The three levels are: the world economy as the network level where services are dispensed; the cities as the nodal level, constituting the knowledge constellations for production of services; and the advanced producer service firms that create and provide the services. Such a triple structure is unusual in network analysis but is by no means unique: in Knoke and Kuklinski's (1982: 16) typology of network relations there is one category with a triple configuration that they call interlocking networks. These are networks wherein the nodes are connected through their constituent subcomponents. A well-known example is the network of corporate boards linked together through people holding overlapping directorships. The directorships example will be used as an initial analogue for the world city network.

In the classic study of overlapping directorships in the United States, Burt (1983) treats directorship ties between firms as a means of lessening market constraints on corporate profits. As non-market connections, these ties reduce the uncertainty of free markets, either through enhanced information or through mutual influences on, even co-ordination of, policy making. The ties of relevance here are the 'direct interlocking ties' where autonomous firms share certain directors (Burt 1983: 74–7). Hence a board will consist of 'inside' and 'outside' directors, and it is the latter who provide the links with other firms to create an interlocking network. This triple configuration consists of a network at the level of the US economy, a nodal level of firms with their boards attempting to reduce market influences, and a sub-nodal level of outside directors linking firms through their board memberships. Burt (ibid.: 85–9) specifies such a network.

However, despite the similarity of form, it is important to recognize the limitations of Burt's network as an analogue of the world city network. Basically, in the analogue network the nodes remain the prime actors in the sense that the firms, through their boards, are the key decision-making unit: they choose (and can dismiss) outside directors and thus control their relations within the network. In contrast, in the world city network as an interlocking network, it is the sub-nodal component, the service firm, that is the prime agency of network production and reproduction. To be sure, city governments will operate in ways to attract and keep leading firms through their 'boosterism' policies and hence may influence relations between nodes, but they are certainly not able to control relations in the way boards of directors do. Hence in the interlocking relations between cities within the world economy, the nodes themselves constitute vital enabling environs to be sure, but, as argued previously, they are not the critical level of decision making within the triple structure. I am aware of no other network specification in which the sub-nodal level is so important.

It follows that if the world city network is indeed a particularly unusual case of an uncommon form of social network, precise specification becomes even more of a prerequisite for advancing researches. The purpose of specification is to make transparent the basic forms in which world city network formation can be described. The idea is to articulate the process in such a way that the unusual and uncommon features of the network do not inhibit analysis in the first instance. Of course, the particular nature of world city network formation will come to the fore in any interpretation of network analysis results, of which more in future chapters.

Initial formal specification

Figure 3.2 has been constructed to aid in describing the formal specification. It depicts a minuscule part of the world city network as an interlocking structure: ten cities (these are the leading world cities, identified as 'alpha' by Beaverstock et al. (1999a)) and three advanced producer service firms, one from advertising, one from finance and one from law. This example will be used in what follows to provide concrete results at different stages of the specification; the results should be treated as strictly illustrative and not as meaningful findings about the world city network.

A universe of m advanced producer service firms located in n world cities is defined. The *elemental attribute* is x_{ij}, where firm j has a presence or not in city i. These simple binary observations can be arrayed as an $n \times m$ presence matrix, $\mathbf{X}$. Normally in this area of research, there is more information than mere presence: for instance, the size of the office of firm j in city i might be available. In general I shall call such extra information on the importance of a city within a firm's office network its *service value*,

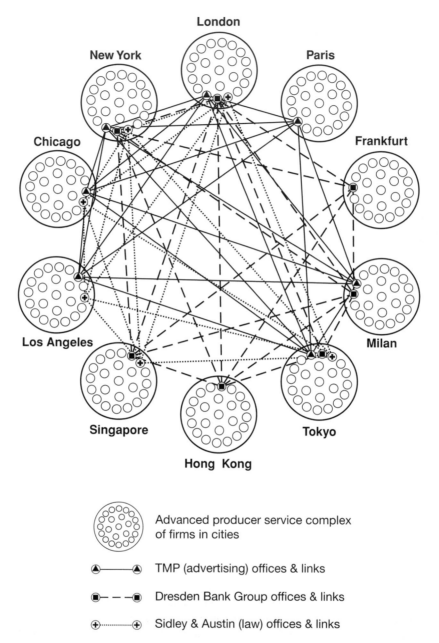

London

New York   Paris

Chicago   Frankfurt

Los Angeles   Milan

Singapore   Tokyo

Hong Kong

Advanced producer service complex of firms in cities

TMP (advertising) offices & links

Dresden Bank Group offices & links

Sidley & Austin (law) offices & links

Figure 3.2 A minuscule portion of the world city network as an interlocking network: ten 'alpha' cities and three advanced producer service firms (from Taylor 2001a)

denoted as v_{ij}. The array of all such values defines a service value matrix, **V**. The conjecture behind using these values is that the larger the office the more connections there are with other offices in a firm's network. For illustrative purposes, a small data set is used with sizes presented as simple integers ranging from 0 to 3: Table 3.1 (using the same firms and cities as Figure 3.2) shows V as an actual 10-city × 3-firm service value matrix.

Table 3.1 *A small service value matrix*

City		Advanced producer service firm			
		I	*II*	*III*	Σ (C_i in eqn 3.3)
1	Chicago	2	0	3	5
2	Frankfurt	0	3	0	3
3	Hong Kong	0	1	0	1
4	London	3	1	3	7
5	Los Angeles	3	0	3	6
6	Milan	1	1	0	2
7	New York	3	2	3	8
8	Paris	1	0	0	1
9	Singapore	0	1	1	2
10	Tokyo	1	1	1	3
	Σ (F_j in eqn 3.2)	14	10	14	(38)

I is TMP (advertising); II is Dresden Banking Group; and III is Sidley & Austin (law)

The first attribute is the *total services in the universe*.

$$S = \sum_i \sum_j v_{ij} \tag{3.1}$$

In Table 3.1 $S = 38$.

The universe attribute from equation 3.1 can be decomposed into two measures of the statuses of either individual firms or cities:

$$F_j = \sum_i v_{ij} \tag{3.2}$$

$$C_i = \sum_j v_{ij} \tag{3.3}$$

Both sets of sums are given in Table 3.1. Equation 3.2 measures the total service value provided by firm F_j across all cities. This defines the *firm service status* of F_j and can be used in comparative analyses to stratify or rank firms: in Table 3.1 the Dresden Banking Group has the lowest rank. Equation 3.3 measures the total service value provided within city C_i by all firms: in Table 3.1 New York has the highest sum. This is a measure of the world city status of C_i that I will call the *site service status* for reasons that will become apparent in what follows. Since I are concerned with city networks, concentration is upon developing the specification by building on equation 3.3, and not equation 3.2, although there is a parallel argument for firms to that which can be presented for cities. However, the site status defines only the importance of a node (city); it says nothing about relations between nodes (cities).

The basic relational element for each pair of cities derived from matrix **V** is

$$r_{ab,j} = v_{aj} \cdot v_{bj} \tag{3.4}$$

which defines the relation between cities a and b in terms of firm j. This is an *elemental interlock* between the two cities for one firm. The aggregate *city interlock* is then produced from

$$r_{ab} = \sum_j r_{ab,j} \tag{3.5}$$

For each city there are $n - 1$ such links; that is, one to every other city. Table 3.2 shows the products from equation 3.4 and the sums from equation 3.5 for the first two cities in Table 3.1, Chicago and Frankfurt. In an egocentric analysis (i.e. taking one city at a time), these links can be used to measure the overall status of a city within the network:

$$N_a = \sum_i r_{ai} \qquad a \neq i \tag{3.6}$$

This will be called the *interlock connectivity*, which defines the *situational status* of city a. From Table 3.2 it can be seen that Chicago's interlock connectivity is 57 and Frankfurt's is 21. The sum of the situational status for all cities defines the *total network interlock*, T:

$$T = \sum_i N_i \tag{3.7}$$

For the cities in Table 3.1, $T = 401$. This sum can be used to express the interlock connectivity of a city as its proportion of total interlock (T):

$$L_a = (N_a/T) \tag{3.8}$$

For Chicago and Frankfurt the proportions are 0.12 and 0.05 respectively, which compare with the highest proportion, New York's $L = 0.2$.

 With large data matrices N_a becomes very large and L_a becomes very small, making both unwieldy empirical measures. As an alternative, a city's interlock connectivity can be defined as a proportion of the highest city connectivity in the universe. Let city h have the highest connectivity; then interlock connectivity can be expressed as the proportion

$$P_a = (N_a/N_h) \tag{3.9}$$

The latter definition yields values ranging from unity to zero (for the cities in Table 3.1 they range from New York scoring 1.0 to Frankfurt scoring 0.25). It is this easily interpreted form of interlock connectivity that will be used in subsequent analyses.

Table 3.2 *Elemental interlock links*

City		For Chicago:				For Frankfurt:			
		I	*II*	*III*	$\Sigma*$	*I*	*II*	*III*	$\Sigma*$
1	Chicago	—	—	—	—	0	0	0	0
2	Frankfurt	0	0	0	0	—	—	—	—
3	Hong Kong	0	0	0	0	0	3	0	3
4	London	6	0	9	15	0	3	0	3
5	Los Angeles	6	0	9	15	0	0	0	0
6	Milan	2	0	0	2	0	3	0	3
7	New York	6	0	9	15	0	6	0	6
8	Paris	2	0	0	2	0	0	0	0
9	Singapore	0	0	3	3	0	3	0	3
10	Tokyo	2	0	3	5	0	3	0	3
	Σ (N_a in eqn 3.6)				57				21

Note: r_{ab} from eqn 3.5 and for matrix V (Table 3.1)

Creating data to describe the world city network

The service matrix **V**, portrayed as *V* in Table 3.1, precisely specifies what data are required to describe the world city network. I need: (a) to identify a set of global service firms; (b) to select a set of cities from within which can be found the cities that make up the world city network; and (c) to find or derive service values that show the importance of each city to the service office network of each firm. The argument proceeds in two stages. First, the process of gathering the appropriate information is described. The method employed is described as 'scavenging', since any information that can inform the data needs is recorded. Second, the conversion of this multifarious information into comparable data across firms is described. The data are produced by devising a uniform scale of service value that is then applied separately to the specific information gathered on each firm. The end result is a data matrix of the service values of global service firms across world cities to describe and analyse the world city network as it existed in the year 2000.

Information gathering

Without recourse to reliance on official published data, the specific collection of a large quantity of information on private corporations is fraught with difficulty. The most obvious problem is confidentiality since, as a general rule, no corporation wants to reveal its strategies, including locational strategies, to its competitors. However, advanced producer service firms are the focus of the information gathering here and they depart from this rule in one crucial respect. These firms provide knowledge-based (expert/ profession/creative) services to other corporations to facilitate their business activities. Such corporate service firms have benefited immensely from the technological advances in computing and communications that have allowed them to broaden the geographical distribution of their service provision. For instance, law firms have been traditionally associated with a particular city and its local client base – a 'New York law firm', a 'Boston law firm', and so on – but under conditions of contemporary globalization a few firms have chosen to pursue a strategy of providing legal services across the world. In such a situation, 'global presence' is an integral part of the firm's public marketing and recruitment policies. For instance, new potential clients from around the world will want to know the geographical range of the services on offer. Also, since these are knowledge-based firms, a global scope is very obviously an important advantage in signing up the best of the next generation of key workers. Hence among producer service firms, locational strategy is perforce quite transparent. Typically, the web sites of such firms provide an option to select 'location', giving addresses of offices, often with a world map of their distribution to emphasize their global presence. Advantage is taken of this geographical transparency for information gathering.

The starting point is to find basic information on where major service firms are present in order to select those firms pursuing a global strategy. On the basis of experience from previous experiments in this field, a firm is deemed to be pursuing a global locational strategy when it has offices in at least fifteen different cities, including one or more cities in each of the prime globalization arenas: northern America (USA and Canada), Western Europe and Pacific Asia (as identified in Beaverstock *et al.* 1999b, 2000b). Selection of firms having met this condition is quite pragmatic. From rankings showing the top firms in different sectors, firms are selected on the basis of the availability of information on their office network. In addition, since one obvious research interest is comparison across different service sectors, firms are included in the data only in sectors for which at least ten firms can be identified. Using these criteria, eighteen accountancy firms, fifteen advertising firms, twenty-three banking/finance firms, eleven insurance firms, sixteen law firms and seventeen management consultancy firms have been selected. These constitute the 'GaWC 100', the global service firms at the heart of this research exercise (see Appendix A).

Although the starting point is firms, the information collected defines networks. Many global service firms exist as 'groups'. For instance, in accountancy there are alliances of medium-sized firms constituted as networks in order to compete globally with the very large firms that lead this sector. In other sectors, take-over activity has led to a corporate structure of core firm plus subsidiaries, with the latter providing distinctive services as an additional dimension to the main service provision – for instance, as the investment arm of a mainstream bank. Sometimes the structure of core firm plus subsidiaries straddles the sector boundary, such as banks owning insurance companies. The latter are treated here within a single network and allocated to the core company's sector. Basically the networks are defined by the worldwide service contacts provided for clients on a firm's web site. Thus the GaWC 100 constitutes a large sample of global service networks.

A few of the larger firms have branches in many hundreds, even thousands, of cities and towns. The data collection has been restricted to the more important cities for two reasons. The first is analytical: the more cities included, the more sparse the final matrix will become, with nearly all the GaWC 100 networks not present in the smaller cities and towns. The second is theoretical: the interest is in the more important inter-city relations, ultimately the world city network. Nevertheless, it is also important not to omit any possible significant node in the world city network, so that a relatively large number of cities need to be selected. Additionally, it is necessary to ensure that all continents are reasonably represented. The final selection of cities is based upon previous experiments and includes the capital cities of all but the smallest states plus numerous other cities of economic importance. The resulting set consists of 315 cities (see Appendix B). It is these cities that are used in recording information on the global service networks of firms.

Selecting firms and cities is relatively straightforward; problems arise when attempts are made to gather information on the importance of a given city to a firm's global service provision. There is no simple, consistent set of information available across firms. The prime sources of information are web sites, and every one is different among the 100 firms. It is necessary to scavenge all possible relevant available information, firm by firm, from these sites and supplement it with material from any other sources available such as annual reports and internal directories. For each firm, two types of information have been gathered. First, information about the *size of a firm's presence* in a city is obtained. Ideally, information on the number of professional practitioners listed as working in the firm's office in a given city is needed. Such information is widely available for law firms but is relatively uncommon in other sectors. Here other information has to be used such as the number of offices the firm has in a city. Second, the *extra-locational functions* of a firm's office in a city are recorded. Headquarters functions are the obvious example, but other features like subsidiary HQs and regional offices are recorded. Any information that informs these two features of a firm's presence in a city is collected in this scavenger method of information gathering. The end result is that for each of the 100 firms, information is available to create service values in each of 315 cities.

Data production

The problem with the scavenger method is that the type and amount of information vary immensely across firms. For instance, some firms have geographical jurisdictions of offices that are 'regional' (transnational) in scope, others have 'national offices', or there may be 'area offices' or 'division offices', with wide variation in the geographical meaning of each category. In addition, many firms will have no specified geographical jurisdictions for any of their offices. Some information is quite straightforward, as when a hierarchical arrangement is shown through contact with an office being routed through an office in another city. But it is more common to find a confusing range of information indicating the special importance of an office. Here is a list of some such designations: 'key offices', 'main branches', 'global offices', 'international offices', 'hub offices', 'major operation offices', 'competence centres' (for a given function), 'asset management centres', 'global investment service centres', offices with 'international trade contacts' or simply with 'international contacts', offices for 'multinational corporate customers', offices housing 'senior managers' or 'senior partners', and offices of 'core firms' within alliances. This is a rich vein of information but much work is required to convert it into usable data to compare firms across cities.

In conversion from information to data there is always a tension between keeping as much of the original material as possible and creating a credible ordering that accommodates all degrees of information across cases. In this exercise there is very detailed information for some firms and much less for others. This tension is resolved here by devising a relatively simple scoring system to accommodate the multifarious information gathered. A six-point service value scale is used where two levels are automatically given: obviously 0 is scored where there is no presence of a firm in a city, and 5 is scored for the city that houses a firm's headquarters. Hence decision making on scoring focuses upon allocating the middle four scores (1, 2, 3 and 4) to describe the service value of a firm in a city. This means that for each firm three boundary lines have to be specified: between 1 and 2, 2 and 3, and 3 and 4.

The basic strategy of allocation is to begin with the assumption that all cities with a non-HQ presence of a firm score 2. This score represents the 'normal' or 'typical' service level for the given firm in a city. To determine such normality requires inspection of the distribution of information across all cities for that firm. To alter this score there has to be a specific reason. For instance, a city where contact with a firm's office is referred elsewhere (i.e. to another office of the firm in a different city) will be allocated a service value score reduced to 1. In other firms where there is full information on numbers of practitioners, a city with an office showing very few (perhaps none) professional practitioners would also score 1. The point is that the boundary between 1 and 2 will differ across firms depending on information available. The same is true of the other boundaries. Generally, the boundary between 2 and 3 has been based upon size factors, and that between 3 and 4 on extra-territorial factors. For instance, exceptionally large offices with many practitioners will lead to a city scoring 3, while location of regional headquarters will lead to a city scoring 4. In practice, size and extra-locational information have been mixed where possible in deciding on the boundaries for each firm. The end result is the service value matrix V, a 315 × 100 data array with v_{ij} ranging from 0 to 5.

How credible are these data? They are far from perfect, largely dependent as they are on what information is available on web sites. But the key issue is the subjectivity inherent in the process of this data creation: the resulting data do not have the key property of inter-subjectivity. That is to say, two people using the same information will not always decide on the same boundaries. Given the nature of the information this is inevitable. One fundamental question arises. Does this issue lead to so much uncertainty in the data that the exercise is irredeemably flawed? There are two answers to counter this concern. First, the means of scoring has been designed to be as simple as possible, pivoting on '2 as normal' and with decision-making limited to just three boundaries. Second, the exercise is carried out over a large number of firms so that particular differences will most likely be ironed out in the aggregate analyses that the data are designed for. Thus I am satisfied that I have produced credible data for describing the world city network in 2000.

Global network connectivity

Measurements of firms and cities in terms of their network locations can be easily derived from V. The sums for columns, rows and the total (equations 3.1–3.3) provide initial description of the universe of global services as defined by the GaWC 100. The total service sum is 16,901 and the top ten firms and cities in terms of quantity of service values are given in Tables 3.3 and 3.4. There is nothing surprising in these rankings, with both tables, coincidently, showing a gap separating the top two from the rest. In Table 3.3 the dominance of the accountancy sector is expected, given the large number of offices the major firms in this sector operate. The cities listed in Table 3.4 are exactly the same as the ten cities designated as alpha world cities in an earlier study based upon different data (Beaverstock *et al.* 1999a). The obvious plausibility of these first simple measurements provides an initial credibility to the new data matrix.

The total service values given in Table 3.3 measure the *site service status* of the cities. This is a measure of the size of cities as service nodes in the world city network. The *situational status* of a city within the network is given by combining equations 3.5 and 3.6:

$$N_a = \sum_i \sum_j v_{aj} \cdot v_{ij} \qquad \text{where } a \neq i \tag{3.10}$$

Table 3.3 *Top ten firms ranked by total service value across 315 cities*

Rank	Firm	Sector	Total
1	KPMG	Accountancy	618
2	PricewaterhouseCoopers	Accountancy	559
3	Arthur Andersen	Accountancy	392
4	CitiGroup	Banking/finance	377
5	Moores Rowland Int.	Accountancy	367
6	HLB International	Accountancy	357
7	BBDO Worldwide	Advertising	351
8	RSM International	Accountancy	346
9	HSBC	Banking/finance	345
10	PFK International	Accountancy	341

Table 3.4 *Top ten cities ranked by total service value across 100 firms*

Rank	City	Total
1	London	368
2	New York	357
3	Hong Kong	253
4	Tokyo	244
5	Paris	235
6	Singapore	229
7	Chicago	213
8	Los Angeles	201
9	Frankfurt	193
10	Milan	191

so that N_a is the nodal connectivity of city a in the network defined as n cities, i, and m firms, j, with v as the service values in $\mathbf{V}$. Given the range and scope of the data used here, this measure can be reasonably designated as the *global network connectivity* of a city. The sum of all these city connectivities is 4,078,256 (from equation 3.7). Individual city values can be expressed as a proportion of this grand total of interlocking connections (equation 3.8) or as a proportion of the highest individual score (equation 3.9).

The top ten cities ranked in terms of global network connectivity are shown as gross and both proportional measures in Table 3.5. Not surprisingly, this table is similar to Table 3.4 but it is not exactly the same: Paris jumps ahead of Tokyo and Milan jumps ahead of Los Angeles, while Frankfurt drops out, to be replaced by Madrid. What this is indicating is that the important firms in the cities that rise in the ranking are relatively more connected than the equivalent firms in cities falling in the rankings – hence the greater global connectivity of, say, Paris over Tokyo. In terms of comparing the relative utilities of the site and situational measures, global connectivity is an aggregate *relational* measure and therefore is the preferred means of assessing the importance of cities in a network context. In addition, the situational status of cities is the more analytically interesting since it leads on to the creation of connectivity matrices and the more sophisticated data analyses of later chapters. In what follows, city connectivities will be expressed in the more convenient proportional form that ranges from 1 to 0 (equation 3.9 and the final column in Table 3.5).

The global network connectivities of all 315 cities are shown in Figure 3.3. This graph provides an initial 'test' of whether these connectivities form a hierarchical structure. It is based on the conventional format for illustrating the structure of 'national urban systems' where the logarithm of city populations is arrayed against the logarithm of their ranks. In national-scale studies a linear plot represents the rank size rule and a 'lazy L-shaped' distribution denotes a primate city pattern (see Figure 1.4, p. 17). As noted in Chapter 1, in effect these two forms of curve denote different types of hierarchical urban systems, the former an orderly integrated hierarchy, the latter a simple pattern of domination. In the one example of using this method to look at the global distribution of city populations (Ettlinger and Archer 1987), the curve takes neither form but is the inverse of the primate city pattern; this indicates a lack of hierarchy in the structure. It is interesting that Figure 3.3 also shows such a curve. Of course, replacing simple city populations by measures of their network connectivity is a far better way of evaluating the nature of the urban pattern under scrutiny. Hence it is initially significant

Table 3.5 Top 10 cities ranked by global network connectivity

Rank	City	Gross connectivity	Proportional connectivity	Proportional to highest
1	London	63,399	0.01556	1.00
2	New York	61,895	0.01552	0.98
3	Hong Kong	44,817	0.01100	0.71
4	Paris	44,323	0.01087	0.70
5	Tokyo	43,781	0.01076	0.69
6	Singapore	40,909	0.01003	0.65
7	Chicago	39,025	0.00957	0.62
8	Milan	38,265	0.00938	0.60
9	Los Angeles	38,009	0.00932	0.60
10	Madrid	37,698	0.00924	0.59

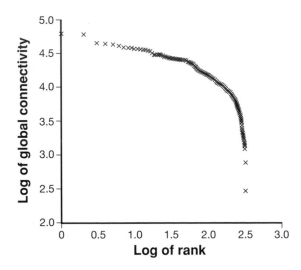

Figure 3.3 Global network connectivities of the 315 cities studied (from Taylor *et al.* 2002a)

that Figure 3.3 depicts a distinctively non-hierarchical urban structure. This is empirical support for the argument that world cities constitute a complex network rather than a simple hierarchy. Although the first two ranks stand out (London and New York), the rest of the curve shows that this is not a 'binary' (or 'double primate') city pattern. There may or may not be hierarchical patterns *within the spatial organization of individual firms* at the global scale (it depends on their particular strategies), but when aggregated, the result is a world city *network*.

◤4◥ Geographies of connectivity

> Cities are primarily focal points of power based upon communication; their power reflects their accessibility – the range and quality of the contacts and relationships that the city has with the rest of the world.
>
> (Knight 1989a: 40)

This is the first largely empirical chapter of the book. In the previous chapter I have specified the world city network and collected data accordingly. This produced a large data matrix from which simple initial results have been presented. Here I begin the task of comprehensively exploring these data. The focus is on interlock connectivities, which I have termed the global network connectivities of cities.

The chapter divides into four parts. The starting point is a cartogram of city connectivities that shows a global-scale archipelago of cities (Figure 4.1) reminiscent of Abu-Lughod's (1989) thirteenth-century transcontinental archipelago described in Chapter 1. But the contemporary version is a 'world city archipelago', a much more intensive single network of cities, as is shown in some detail in what follows. But first, I explore the pattern of world cities across regions and also consider the holes in the archipelago, regions 'beyond world cities'. The next two sections present results from disaggregating the connectivities. In the second section connectivities are divided into sector components that show cities as different types of service centre. The third section looks at the service values of cities and suggests different levels and types of power in the network. In the final section I take advantage of the fact that the network methodology need not be limited to service providers. In this section the global service connectivities are compared to other connectivities of world cities.

Results are shown on a cartogram illustrating the most connected cities (Figure 4.1) because this mode of presentation solves the problem of depicting an uneven distribution of cities across the world. City concentration in some regions (e.g. in Western Europe) coupled with sparseness of cities elsewhere (e.g. in Africa) makes depiction of results on orthodox maps, with extremes of overlaps and empty spaces, sometimes difficult to perceive and interpret. Hence the cartogram, wherein each city is given its own equal space in approximately its correct geographical position. I have had to limit the number of cities to aid comprehension of the cartogram and so as not to lose sight of the leading cities across the world. I define this 'operational roster of world cities' as those with at least one-fifth of London's connectivity. This cut-off point is purely arbitrary: remember, the graph in Figure 3.3 is smooth from rank 3 downwards. It has been chosen, first, because it gives a reasonably large number of cities – 123 is much larger than in other world city studies – and second, because it provides a reasonable coverage of most world regions. Africa is represented by six cities, including two inter-tropical cities, and the only regions not included are Central America (nothing between Mexico City and Panama City) and Central Asia (nothing between Moscow and Beijing).

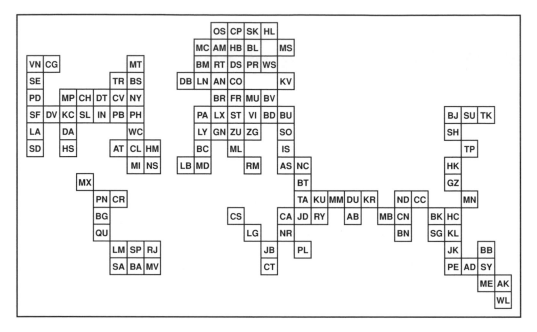

Figure 4.1 An archipelago of world cities. The cartogram places cities in their approximate relative geographical positions. The codes for cities are as follows: AB Abu Dubai; AD Adelaide; AK Auckland; AM Amsterdam; AN Antwerp; AS Athens; AT Atlanta; BA Buenos Aires; BB Brisbane; BC Barcelona; BD Budapest; BG Bogotá; BJ Beijing; BK Bangkok; BL Berlin; BM Birmingham; BN Bangalore; BR Brussels; BS Boston; BT Beirut; BU Bucharest; BV Bratislava; CA Cairo; CC Calcutta; CG Calgary; CH Chicago; CL Charlotte; CN Chennai; CO Cologne; CP Copenhagen; CR Caracas; CS Casablanca; CT Cape Town; CV Cleveland; DA Dallas; DB Dublin; DS Düsseldorf; DT Detroit; DU Dubai; DV Denver; FR Frankfurt; GN Geneva; GZ Guangzhou; HB Hamburg; HC Ho Chi Minh City; HK Hong Kong; HL Helsinki; HM Hamilton (Bermuda); HS Houston; IN Indianapolis; IS Istanbul; JB Johannesburg; JD Jeddah; JK Jakarta; KC Kansas City; KL Kuala Lumpur; KR Karachi; KU Kuwait; KV Kiev; LA Los Angeles; LB Lisbon; LG Lagos; LM Lima; LN London; LX Luxembourg City; LY Lyons; MB Mumbai; MC Manchester; MD Madrid; ME Melbourne; MI Miami; ML Milan; MM Manama; MN Manila; MP Minneapolis; MS Moscow; MT Montreal; MU Munich; MX Mexico City; NC Nicosia; ND New Delhi; NR Nairobi; NS Nassau; NY New York; OS Oslo; PA Paris; PB Pittsburgh; PD Portland; PE Perth; PH Philadelphia; PL Port Louis; PN Panama City; PR Prague; QU Quito; RJ Rio de Janeiro; RM Rome; RT Rotterdam; RY Riyadh; SA Santiago; SD San Diego; SE Seattle; SF San Francisco; SG Singapore; SH Shanghai; SK Stockholm; SL St Louis; SO Sofia; SP São Paulo; ST Stuttgart; SU Seoul; SY Sydney; TA Tel Aviv; TP Taipei; TR Toronto; TY Tokyo; VI Vienna; VN Vancouver; WC Washington, DC; WL Wellington; WS Warsaw; ZG Zagreb; ZU Zurich.

Figure 4.1 shows a great global archipelago of cities, and therefore the first geographical result of the book is that there is indeed a worldwide pattern of global service centres, albeit an uneven one.

The geography of global network connectivity

The geography of city connectivities is depicted in Figure 4.2. The unevenness in the distribution of world cities is exacerbated by the pattern of relative levels of global network connectivity. At its simplest, the cartogram reproduces the old 'North–South' divide: higher-connected cities tend to be in the 'North' and lower-connected cities in the 'South', with the western Pacific Rim firmly bucking this trend. But, of course, it is much more complicated; this simple, not to say simplistic, interpretation is only a trend, with many lower-connectivity cities in the 'North' and some higher connectivity in the 'South' beyond the Pacific. To explicate this geography I will describe the pattern in more detail at different scales.

Regional contrasts

If we move now from 'North–South' terminology to world-systems language, Figure 4.2 illustrates clearly the three contemporary zones of the core of the world-economy: northern America, Western Europe and parts of Pacific Asia. However, this is not a homogeneous core: the three zones have very different histories associated with their trajectories to core status and this is reflected in Figure 4.2.

The oldest, indeed original, core zone is Western Europe, and this is reflected in two features. First, this region has more world cities (32) in Figure 4.2 than the other regions, and second, there is a wide range of levels of connectedness among the region's cities. In other words, in this region there are a variety of cities of varying importance all linking into the world city network. This is the complete opposite of Pacific Asia, in which the connectivity levels of the cities is generally top-heavy. As this region is the most recent of the core zones, nearly all its less important cities have not made the threshold for the world city network as defined in Figure 4.2. Thus this region has far fewer world cities (13) than Western Europe, although the number increases to twenty if we add Australasian cities to create a Western Pacific Rim region. The third core zone, northern America (i.e. the United States and Canada), is in between the other two historically and in numbers of world cities identified (27). However, in this case the

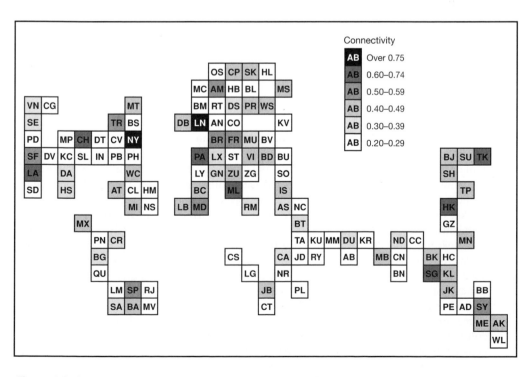

Figure 4.2 Global network connectivity (from Taylor *et al.* 2002c). (For city codes, see Figure 4.1.)

range of levels of connectedness is very similar to that of Western Europe, with numerous less important cities joining the world city network. But there is a difference: in northern America the more connected cities tend to be in the east and west of the region, leaving the centre bereft of well-connected cities apart from the major exception of Chicago.

Beyond the core there are no regions with any concentration of highly connected cities. The most common pattern is for capital cities to take on the world city role. In Eastern Europe (the former Communist states) this is most certainly the case: the only cities (8) that feature are the capital cities of the larger states. Having lost its political and economic distinctiveness, this region has become an appendage to the West European core. The same may be said for Latin America (11 cities) with respect to northern America, where again capital cities dominate, although in this case São Paulo, despite being neither former nor current Brazilian capital city, has become a highly connected world city in its own right. This pattern is similar in South Asia, where Mumbai, another non-capital, has become a highly connected world city. In contrast, the large North African/West Asian region (11 cities) has no such well-connected world city; the only cities that appear as possible candidates for becoming a regional focus are Cairo, Beirut and Dubai. Sub-Saharan Africa has only four cities but it does sport a clear regional leading city in terms of connectivity: Johannesburg.

What does all this locational detail mean? First, uneven globalization has spawned an uneven world city distribution but not a simple one. The number of cities featuring outside the core is perhaps surprising. Whether these represent simple continuations of colonial and post-colonial 'economic sinks' or genuine components to a network is an issue that I address later in the chapter. The most interesting region is Pacific Asia. The latest core zone, it is like a non-core region but with its leading cities upped in terms of connectivity, leaving few low-connectivity cities. The lack of the latter suggests that numerous small parts of Pacific Asia remain poorly connected to the world city network. This may be the mark of a region in transition between core and semi-peripheral status. If this is the case, then it will be necessary to look for increased global network connectivity for some of the region's lesser cities as a future sign of consolidation of core status.

National differences

Although I am working on a model of world-economy constituted by city economies, this does not mean that all markets operate at just these two scales. The idea of a 'national economy' may be a myth but, as shown in Chapter 2, this does not mean that there are no national market effects on cities as service centres. States have been and continue to be powerful shapers of markets if not creators of economies. In terms of the world market of business services, states are anything but irrelevant to world cities and their connectivities.

States affect different services in different ways. For the various financial services there are regulations whose level of control varies by country. For law, states constitute legal jurisdictions that have to be coped with in any transnational commercial project. States also legitimate professional gatekeepers: who can and who cannot practise law, and other professions, in their territory. For advertising and management consultancy, states are less intrusive but here other national effects become important. These are cultural effects on how products will be received. Global advertising has to deal with consumers who not only speak different languages in different countries, but may also have very different reactions to similar translated language or visual signals. Global management consultancy has to cope with many business mores; paternalistic companies where management merely means 'direction' provide a common challenge. The point of all these examples is to reinforce the idea broached in Chapter 2 that even in the world of advanced producer services the national space of places cannot be ignored: cities as nodes in the global space of flows are also cities within countries.

As described in Chapter 2, Jacobs (1984) posits national urban development processes that favour one city over all others in a country. Such a process provides that city with a particularly strong platform on which to globalize. This will be especially the case as new firms begin a global strategy and plan to serve national markets through just a single office. Hill and Fujita (1995) have referred to 'Osaka's Tokyo problem', but it is clearly much more than a Japanese phenomenon. As well as the Japanese market being largely serviced through Tokyo, the Austrian market can be served through Vienna, the Swedish market through Stockholm, the UK market through London, and so on. Thus the primate city of 'national urban systems' become the 'national world city', a national gateway into and out of the world market for services.

This process can be explored through computing the ratio of global network connectivities between the city with the highest level in a country and the city ranked second. These are shown for a selection of twenty-five large countries in Table 4.1. The countries are presented as two groups using a ratio of 2 as the divider, i.e. whether the leading city in a country is more or less than two times more connected than its closest rival. This value has resonance with national-level urban studies as specifying the rank-size rule (see Chapter 1), with values above 2 indicating different levels of primacy. A slight majority (14) of countries show a connectivity primacy. In all but one case it is the capital city that has the high connectivity, and the exception is an ex-capital (Istanbul/ Turkey). The ratios vary from Vienna/Linz to Amsterdam/Rotterdam but all indicate a dominating world city linking its national market to the world market. In contrast, there are eight countries that appear not to have primate tendencies in terms of world city connectivities. These all have one or more of the following characteristics: large size, decentralized polity, multiple cultures. These characteristics are precisely the opposite of that found in the connectivity primate city states: mainly small countries plus a few larger countries historically notorious for their political centralization: the United Kingdom, France, Japan, Mexico and Russia.

Table 4.1 *The connectivity ratios between the top two cities for selected countries*

Country	Ratio	Country	Ratio
Austria	8.18	Brazil	1.87
Turkey	7.92	South Africa	1.73
South Korea	6.48	New Zealand	1.68
Egypt	5.80	Italy	1.66
Denmark	5.43	China	1.65
Britain	4.44	USA	1.59
Japan	3.72	Switzerland	1.56
Colombia	3.20	Canada	1.49
Sweden	3.16	Germany	1.44
France	2.89	Spain	1.39
Mexico	2.64	India	1.31
Russia	2.45		
Belgium	2.35		
Netherlands	2.20		

The leading two cities in a country do not tell the whole story of how a national market links to the world market. Obviously, the low ratios could indicate both a 'dual primate' pattern and a smooth hierarchical sequence. To distinguish between these and other possibilities, Figures 4.3 and 4.4 show graphs of the top five cities in terms of connectivity for twelve countries. These are separated into two groups on the basis of size: I term them nation-states and continental states. In the first group (Figure 4.3) the connectivity primacy of the United Kingdom, France and Japan is confirmed and Italy and Spain are shown to have dual primacy patterns (Milan–Rome, Madrid–Barcelona). The interesting case is Germany, whose cities form a quite flat distribution showing an almost total lack of primate tendencies. The graphs for the continental states (Figure 4.4) are generally flatter, with Brazil revealed as a dual primate pattern (São Paulo–Rio de Janeiro) and China showing a 'tri-primate' pattern (Hong Kong–Shanghai–Beijing). In general, these are large states that require more than one world city to service subnational regions that are themselves commonly larger than most other nation-states.

In conclusion: the evidence clearly shows that the nature of states influences the nature of a national market's city connections to the world market. The world city network operates with, through and alongside the mosaic of states as well as across them.

Beyond world cities

Despite the worldwide nature of the world city network there are regions where world cities are either sparse or absent. Since the data allow for connectivities to be computed

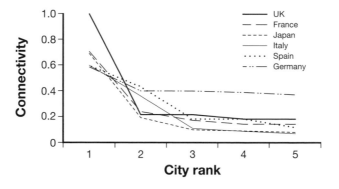

Figure 4.3 City connectivities in nation-states

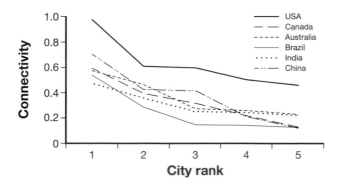

Figure 4.4 City connectivities in continental states

for a total of 315 cities, regions beyond or weakly connected to the world city network as portrayed in Figures 4.1 and 4.2 can be investigated. Thus in what follows, some cities below the top 123 are discussed for the first time.

Sub-Saharan Africa dominates the 'weakly connected' category: there are only four cities that appear in the top 123 cities but there are many more with lower connectivities, as Table 4.2 shows. Generally, apart from smaller South African cities, these lower-connected cities are capital cities, with their ordering approximating the size of their respective national markets: Abidjan (Ivory Coast), Accra (Ghana) in West Africa and Harare (Zimbabwe) and Lusaka (Zambia) in South-Central Africa are the leading cities beyond the world city network of 123 cities. Central Asia is an example of a region where world cities are conspicuous by their absence: Table 4.3 shows the global network connectivities of cities. As with sub-Saharan Africa, in this region featured cities are capital cities ordered with respect to the size of their national markets: Almaty, the

Table 4.2 *The global network connectivities (GNCs) of sub-Saharan African cities*

City	GNC
Johannesburg	0.414
Cape Town	0.239
Nairobi	0.226
Lagos	0.197
Abidjan	0.181
Harare	0.179
Accra	0.167
Lusaka	0.162
Durban	0.151
Windhoek	0.146
Kampala	0.142
Doha	0.140
Dar es Salaam	0.128
Maputo	0.122
Dakar	0.116
Doula	0.105
Gaborone	0.103
Luanda	0.091
Pretoria	0.089
Addis Ababa	0.068
Bulawayo	0.066
Kinshasa	0.049
Mombasa	0.049
Freetown	0.041
Lomé	0.033
Yaoundé	0.024
Monrovia	0.022
Conakry	0.019
Djibouti	0.012
Brazzaville	0.005

Table 4.3 *Global network connectivities (GNCs) of cities in Central Asia*

City	GNC
Almaty	0.173
Baku	0.112
Tashkent	0.099
Yerevan	0.036
Tbilisi	0.033
Ulan Bator	0.021
Kabul	0.005

capital of the largest Central Asian republic, Kazakhstan, is the most connected of the region. What these two examples show is that even where world cities are sparse or absent there are still global service connections into the world city network. I explore this theme in some detail through another region beyond world cities: Central America.

Global service firms vary greatly in their particular global strategies of office location. Smaller firms have obviously to concentrate their resources on a lower number of cities, but the critical determinant of location policy seems to be the service sector. The greatest contrast is between global accountancy firms, which tend to locate in many hundreds of cities, and global law firms, which are usually found only in a select number of world cities. There are five Central American cities in the data (all capital cities again) and each has some direct connections into the world city network through those firms that have an extensive office location policy. In all, there are seventy-nine presences of firms in the GaWC 100 within this region. These are distributed across services and cities in Table 4.4, where it can be seen that San José has most (20) and Managua least (11) presences of the 100 global service firms that constitute the data. Most of the firms with Central American offices are in accountancy and advertising. In aggregate, these firms produce the global network connectivities also shown in Table 4.4. In this column the ordering changes slightly, with Guatemala City having more connectivity despite having fewer firms present than San José. The key point, however, is that, in some sectors at least, Central American cities are part of the office networks that create the world city network. In other words, Table 4.4 confirms that the world city network is not constituted as an exclusionary club of the major cities but has numerous linkages into regions beyond world cities.

By using new data, the argument can be taken further by identifying the cities in the world city network that provide the key linkages for Central America (Brown *et al.* 2002). The obverse of global firms in local cities is the linkages of local firms into world cities. Small non-global firms can operate beyond their normal geographical range by

Table 4.4 *Presence of firms and global network connectivities (GNCs) for Central American cities*

City	AC	AD	BF	IN	LW	MC	Total	GNC
Guatemala City	9	8	1	1	0	0	19	0.181
San José	10	5	3	1	0	1	20	0.175
San Salvador	8	5	3	0	0	1	17	0.165
Tegucigalpa	6	4	1	0	0	1	12	0.130
Managua	4	5	0	1	0	1	11	0.100

Note: AC = accountancy firms; AD = advertising firms; BF = banking/finance firms; IN = insurance firms; LW = law firms; MC = management consultancy firms forming alliances or having other, similar relationships with firms in other regions. This has been a common practice in the banking sector, where 'correspondence banks' are designated. Where a local bank has a client doing business in another area where it does not have an office, it will advise and facilitate that financial service being undertaken by its correspondence bank in that area. This is not a formal alliance but indicates a 'partner of choice' for the mutual benefit of both: the correspondent bank gets the extra business, the local bank does not lose its client through failure to provide adequate geographical scope of service. This does not constitute the ideal of a seamless service under one brand but it does provide the opportunity for local firms to service extra-regional business.

Using information on the correspondent links of twenty-two local Central American banks, a total of 319 links were found of which 168 were with one city: Miami. That over half (53 per cent) of the correspondent links go to banks in just one world city is quite a remarkable finding, a stark indication of the domination of Miami in the external financial connections of Central America. This is a contemporary manifestation of Foucher's (1987: 121) designation of Miami as the 'capital' of the 'American Mediterranean' in the Reagan era. Far behind in second place, and reinforcing the United States' linkage dominance, is the Americas' prime international financial centre, New York, with thirty-five (11 per cent) correspondent links. Although there is this concentration of linkages, Central American correspondence banks are to be found across the world in thirty-four different cities. These are shown in Figure 4.5, which, as well as emphasizing Miami's primacy, has other interesting features, notably the relative importance of Hamburg and Frankfurt, the top two-ranking European cities, with more links than both Madrid, with its colonial/language connections, and London, Europe's prime international financial centre. The relatively low level of connections with Latin America, excepting the two 'neighbour' cities of Mexico City and Panama City, and the lack of importance of Pacific Asian cities are also noteworthy. On the basis of this specific financial link, the conclusion is that Central America has widespread indirect links into the world city network but that these are hugely dominated by its connections to Miami.

Global services across the network

The global network connectivity of a city can be disaggregated into constituent parts in two ways. The most straightforward partition is by service sector: how much of a city's connectivity is due to firms in each of the six sectors. A more subtle partition uses the service values of cities to ascertain power relations. The latter is the subject of the following section; here I focus upon the contributions of different sectors to the global network connectivities of cities.

It is known from Table 3.3 that the business services I am dealing with are by no means equal in their contributions to global network connectivity. As previously noted, accountancy, in particular, has firms that cover many hundreds, indeed sometimes thousands, of cities and therefore it is ubiquitous throughout the top 123 cities. In Table 4.5 the number of cities that are connected through each of the six services are shown. As well as the maximum number of 123 recorded for accountancy, banking/finance also connects every one of our world cities. There are just a small number of cities not connected through advertising, insurance and management consultancy, but quite a few that have no global law firms. Law is the service with by far the lowest presence in cities, indicating its concentration in the more important world cities. Table 4.5 also shows the average percentage of connectivity accounted for by a service across all 123

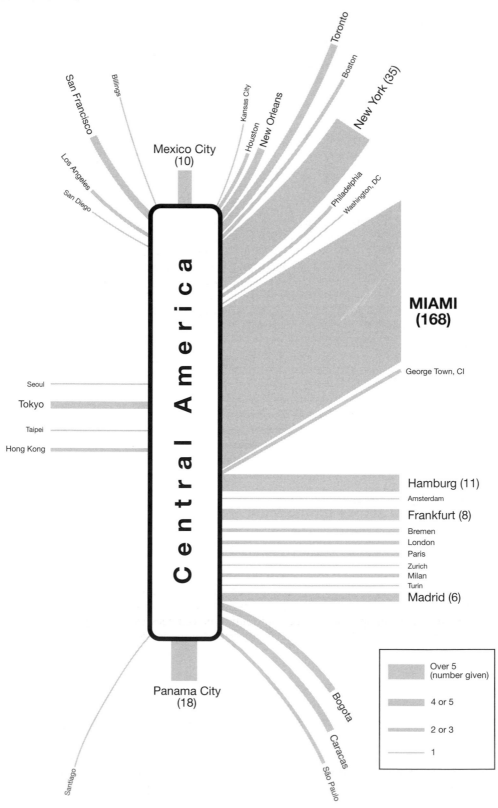

Figure 4.5 Correspondence banks for Central America (from Brown *et al.* 2002)

Table 4.5 *Service sectors, cities and connectivities*

Sector	Number of cities connected through sector	Average connectivity accounted for by sector (%)
Accountancy	123	44.0
Advertising	119	15.7
Banking/finance	123	22.7
Insurance	120	7.8
Law	76	1.7
Management consultancy	119	8.2

cities. Here the importance of accountancy to the global network connectivities is clearly illustrated. This reflects the many offices beyond the top 123 cities. Banking/finance and advertising are both also important contributors. Relatively less important are management consultancy, insurance and, especially, law. With a less than 2 per cent contribution it might be thought that law is almost irrelevant to the global connectivity measures. This inference is not wholly correct: because of the distribution of law offices concentrated in just the leading cities, it contributes in important ways in differentiating cities. This is especially important in the analyses in Part III, but it also shows up in comparing city connectivities later in this chapter.

The sectors can therefore be divided into higher and lower contributors to global network connectivity. They are discussed in order of contribution within these two groups below. To facilitate comparison, each sector is mapped on to the archipelago in the same way using just two categories: cities ranking in the top twenty for a contribution by a sector, and the remaining cities with above-average contribution by a sector.

Before the different sectors are described in detail, one important point needs to be made. Consider the fact that the city that records the highest proportion of its connectivity due to banking/finance is Manama. This city is certainly an important financial centre in the Middle East but it pales in comparison to the level of financial services provided in London and New York. Why do the latter pair rank 57th and 56th respectively on the proportion of their connectivities contributed to by banking/finance? Why not first and second? London and New York are most definitely the top two international financial centres, as analysis in the next section will show, but this is not what is being measured here. Both London and New York are 'well-rounded' global service centres offering much more than banking and financial services. This cannot be said for Manama, which is a regional international financial centre but with few other global services. Thus Manama, and other relatively specialist service centres, will beat London and New York in specific service contribution percentages even though in all sectors London and New York are the most important sites for service providers.

High-contribution sectors

While accountancy, banking/finance and advertising all contribute highly to global network connectivity, the patterns of their contributions are remarkably different.

Accountancy

Although accountancy has been introduced as the most ubiquitous of business services, its geography is much more interesting than this would suggest.

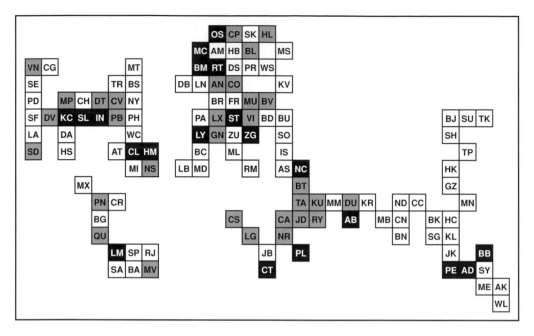

Figure 4.6a Specialist service cities: accountancy. (For city codes, see Figure 4.1.)

The accountancy cartogram (Figure 4.6a) is distinctive in a regional absence: Pacific Asian cities (and South Asian cities) all have below-average accountancy contributions to their global network connectivities. This is the only sector distribution in which one of the three core zones of the world-economy is not represented at least once above average level. Quite clearly, global accountancy has a pronounced 'Western' bias. But within this regional bias there is another clear feature: it is lower-connectivity cities that stand out. Thus the top twenty is a roll-call of lesser world cities through all other regions of the world. The Australian pattern is a good analogue for the rest of the West, with Perth, Brisbane and Adelaide appearing in the top twenty, while Sydney and Melbourne are below average.

In conclusion: it is the less important cities in Western regions (including Western Asia) that are the most dependent on accountancy, the most ubiquitous service, for their global network connectivity.

Banking/finance

Banking/finance is clearly the strategic business service and therefore its geography is particularly interesting.

The banking/finance cartogram (Figure 4.6b) is in important ways the obverse of the accountancy pattern: here Pacific Asia dominates the distribution, with every city above average and, even more impressive, contributing fully half of the top twenty. Otherwise, German cities dominate in Europe with, in addition, the more specialist financial centres of Luxembourg and Geneva, and Manama beyond Europe. Although not absent, the United States is not greatly featured on this cartogram; many of its cities beyond the Pacific coast are not very dependent on banking/finance for their global network connectivities.

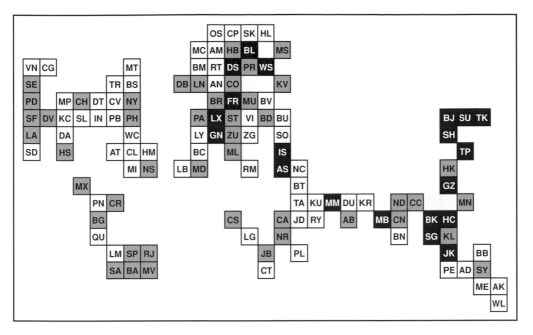

Figure 4.6b Specialist service cities: banking/finance. (For city codes, see Figure 4.1.)

In conclusion: the three core zones define a gradient in city dependences on banking/finance for global network connectivities: cities of the Pacific Asian region are pre-eminently dependent; many European cities are highly dependent; US and Canadian cities have only moderate to low dependence.

Advertising

Global advertising is, in many ways, the epitome of globalization as top brands in many production sectors – cars, oil/petroleum, clothes, food and drink, leisure/vacation products, etc. – are marketed worldwide under single brand names. But this should not be interpreted as necessarily indicating the rise of a homogeneous world market. Rather, there remains a highly fragmented market.

The advertising cartogram (Figure 4.6c) has an unusual pattern that is, at first, surprising. To begin with, the font of consumerism, the United States, is poorly represented, as is Western Europe, consumerism's 'second home'. In fact, nearly all the cities that feature in the top twenty are less-connected world cities outside the core zones. The clue is that they are capital cities or the leading city of a country. In short, these cities are the national media centres, the focus of national advertising markets, notably the homes of national television stations. The exceptions are in the larger countries, where there are regional markets and therefore room for more than one centre for advertising (e.g. in India and China). As well as being poorly represented, the United States is unusual on this cartogram for having only two cities featured within the top twenty: Miami, the 'capital' of much of Latin America, and Detroit, 'Motown', which features a concentration of critical advertising clients.

In conclusion: the geography of connectivity dependence on global advertising is highly nationalized in most of the world; it is the classic case of the international mosaic of territories intersecting with a global space of flows.

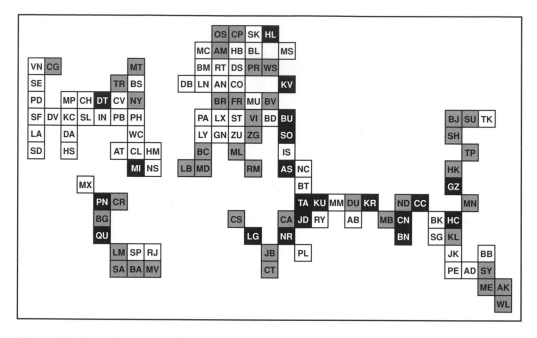

Figure 4.6c Specialist service cities: advertising. (For city codes, see Figure 4.1.)

Lower-contribution sectors

The three sectors contributing less to global network connectivities – management consultancy, insurance and law – are similar in one respect. With fewer connections, those connections are more concentrated in the core zones of the world-economy than for the three higher-contributing sectors.

Management consultancy

Management consultancy is *the* American business service, and therefore it would be expected that US cities would feature particularly strongly in this sector. But the pattern turns out to be more complicated than this.

The management consultancy cartogram (Figure 4.6d) has large concentrations of cities in northern America and Western Europe that are relatively dependent on management consultancy for their global network connectivities. In addition, there are other, small concentrations of western Pacific cities similarly dependent. In the peripheral zones only Latin America has any significant showings. Within the cores zones there does not seem to be any pattern relating to levels of city connectivities, especially in the United States. However, more generally, in northern America there does appear to be a geographical sectional effect, with all but one of the top twenty cities located in the east of the region.

In conclusion: the geography of cities particularly dependent on management consultancy is a patchy one across core zones but with a discernible 'nucleus' in the eastern section of northern America.

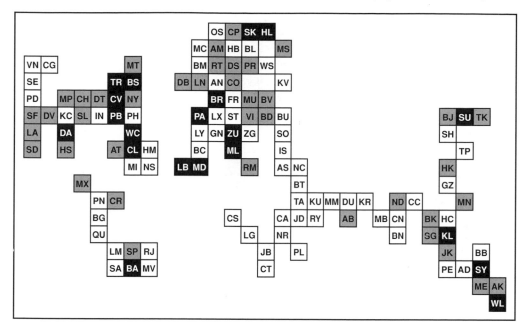

Figure 4.6d Specialist service cities: management consultancy. (For city codes, see Figure 4.1.)

Insurance

Insurance is a specialist financial service that might be expected to be found especially in international financial centres.

The insurance cartogram (Figure 4.6e) has a regional focus similar to that of the management consultancy pattern but with different details within the regions. Thus, there is the same core-zone ordering of where cities are most dependent on insurance: first northern America, followed by Western Europe and then Pacific Asia. The two main regional differences with Figures 4.6c and 4.6d are that Pacific Asian cities are more prominent and Eastern Europe has no showings. There are international financial centres featuring prominently, notably Hong Kong, Luxembourg, Geneva and Hamilton (Bermuda), but the main characteristic is the tendency for lower-connected cities to have their global network connectivity heavily dependent on insurance. This is particularly a feature in the United States.

In conclusion: this is a sector that is primarily important for the global network connectivities of less connected world cities in the two western zones of the world-economy with specific focus on finance centres.

Law

By far the smallest of our six sectors, law is known to be the most concentrated service among world cities.

The law cartogram (Figure 4.6f) has the simplest geography. There are two elements. First, there is concentration but it is not primarily regional; it shows a strong focus on the most connected world cities. The top twenty cities in this case includes a roll-call of leading world cities: London, New York, Hong Kong, Paris, Tokyo, Singapore,

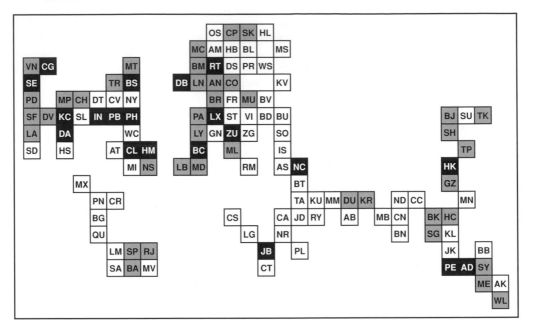

Figure 4.6e Specialist service cities: insurance. (For city codes, see Figure 4.1.)

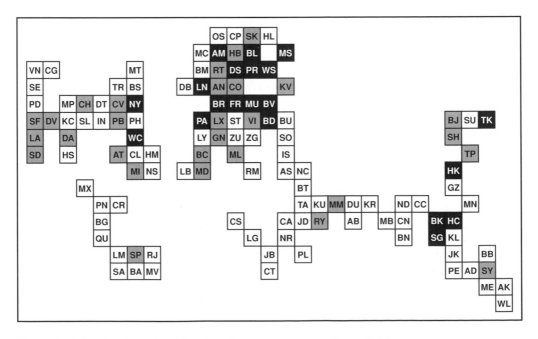

Figure 4.6f Specialist service cities: law. (For city codes, see Figure 4.1.)

Amsterdam, Frankfurt and Washington, DC. Apart from New York and Washington, US cities are less prominent, indicating the dominance of these two cities in global law in the United States. This leaves Europe as the leading zone for the importance of law connectivities. Note the reference to Europe and not just its western section. This is the second feature of the geography: the surprising importance of East European cities. Law was the last of the six services to globalize, and much of this expansion occurred in the 1990s just as Eastern Europe was forming new markets through privatization of state assets. Many of those taking advantage of this new 'frontier' economic bonanza needed transnational legal advice in a legal vacuum: hence the new global strategies of several law firms included this opportune opening.

In conclusion: law is the sector most concentrated in the leading world cities and therefore it is here that it makes its most important contributions to global network connectivities. Exceptions to this rule occur in Eastern Europe for contingent opportunistic reasons.

Power in the network

In the studies of national urban systems reviewed in Chapter 1, the depiction of city hierarchies implied the existence of power relations, but this was hardly ever fully acknowledged. For instance, in key texts such as Bourne's (1975) *Urban Systems* and Johnston's (1982) *The American Urban System*, power is conspicuous by its absence from their respective indexes. However, there is a major exception within this research tradition: the work of John Friedmann (1978). In his 'Spatial organization of power in the development of urban systems' he investigates the effect of both governmental and economic power on the growth of hierarchical urban systems. This is, of course, particularly pertinent to my concerns here because of this author's subsequent pioneering writings on world cities (Friedmann 1986) reviewed in Chapter 1. In the latter, his earlier concern for the spatial organization of power is transferred from the national scale to a global scale where economic power predominates. Thus he identifies the 'global control functions of cities' that constitute a world city hierarchy. It is the purpose of this section to return to Friedmann's original focus on power and to measure contemporary power relativities across world cities.

In devising his world city hierarchy, Friedmann (1986) treated power in the same way as he had done in his earlier national-scale research: as a 'stock of resources' (Friedmann 1978: 329) to be used instrumentally as 'power over' others. This is what Allen (1997: 60) calls 'power as a capacity – a "centred" conception' that he sees as dominating the world cities literature in which cities are centres of control and command (Friedmann 1986: 71). However, power as a capacity is just one of the conceptions of power that Allen (1997) identifies. Instead of this 'nodal' emphasis, more networked conceptions of power can be identified. He notes that Sassen's (1991) conception of the 'global city' recognizes the limitations of the simple capacity conception (Allen 1997: 70) but he finds little evidence for an alternative conception of power in her work. It is there: Sassen (1994) treats her global cities as 'strategic places', a concept that implies much more than simply 'power over'. It seems to me that this is very close to what Allen calls 'power as a medium – a "networked" conception'. Sassen (2000: 148–9) describes a new 'geography of politics' involving 'strategic places . . . bound to each other by the dynamics of economic globalization'. There is emerging 'a transnational urban system' with inter-city relations that transcend simple competition (ibid.: 151). The essence of this is 'power to' rather than 'power over', specifically the power to attract service firms for servicing global capital. This global centring of power in cities is less hierarchical in nature and more networked. In a network, power is much more diffuse as every node has a particular niche that is part of the reproduction of the whole. In other words, complementary relations are more important than competitive ones (Powell 1990). This means that every city, as a node in an urban network, embodies an incipient power of position.

Because Sassen (1994) focuses on 'centrality', Allen (1999) identifies Castells (1996), with his concept of a 'space of flows', as better describing network power among world cities. As I have shown previously, for Castells (1996), world cities are not simply places, they are processes, hubs through which flows are articulated, with power residing in the flows themselves. Thus Allen (1999: 202–3) sets up an opposition of 'city networks' versus 'networks of cities'; that is to say, whether the cities 'run' the networks (Sassen) or the networks 'generate' the cities (Castells). This stark contrast is good for highlighting key issues in the literature but, as Allen (1999: 203) admits, 'probably overstates the differences'.

I will not choose between these alternative loci of power by level. In fact, I suspect that this is a theoretical nicety that cannot be resolved empirically. Thus I will not be attempting to distinguish the nodes from the flows in this power analysis, but I will be focusing on relations between world cities as the basis of their power, however conceived. An eclectic theoretical position with respect to conceptions of power is taken: both capacity/command and medium/network conceptions of power are incorporated in the analyses. A reading of power is attempted within the world city network as both a capacity expressing hierarchical tendencies and a collective medium with differences in power expressed through position in the network.

Control and command centres

Starting with Friedmann's (1986) original conception of world cities as 'control and command centres'. I explore this idea empirically in two ways that are termed domination and control. Both concepts are based upon asymmetric relations: *domination* is taken to mean a more general expression of power through dissecting connectivity, and *command* involves organization through actual direction from above.

Dominant centres

In computing the global network connectivity of a city, its service values are multiplied by the service value of each other city for a given firm (equation 3.6). For each city these products can be classified into three types. Where the city in question has the higher service value it can be referred to as a *dominant* connection; where it has the lower value it is a *subordinate* connection; otherwise, where both values are the same, there is neither dominance nor subordination. From this I can dissect the network connectivity of a city into three parts: *connectivity-through-dominance, connectivity-through-subordination* and *neutral connectivity*. In such an analysis only thirty-four cities have more connectivity-through-dominance than connectivity-through-subordination. These are shown in Figure 4.7, where they are differentiated by the ratio between the two types of connectivity. London and New York stand out with 17.5 and 14.9 times more domination than subordination in their network connectivities respectively. These are designated 'mega' dominant centres, given that the next highest ratio is only 3.5. There are six cities with ratios from 2.5 to 3.5 and they are designated 'major' dominant centres in Figure 4.7. Given their overall importance in connectivity, it is not surprising that Hong Kong, Paris, Tokyo and Chicago appear in this category, but Frankfurt and Miami are less expected. Frankfurt actually ranks fourth, a position that

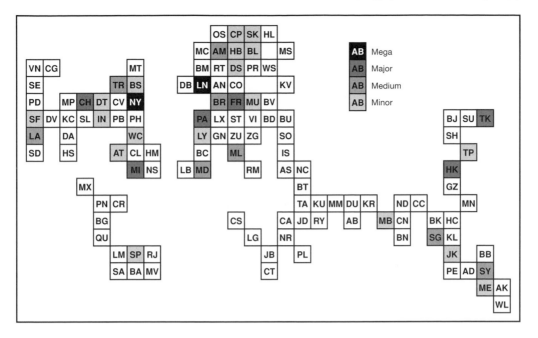

Figure 4.7 Dominant cities (from Taylor *et al.* 2002c). (For city codes, see Figure 4.1.)

is far above its usual world city ranking. This suggests that its role as EU financial centre is reflected as major dominance. Miami's major dominance status is obviously related to its regional functions, to be detailed further in what follows. The cut-off point between 'medium' and 'minor' dominant centres is 1.5. The most notable feature of the distribution of cities in Figure 4.7 compared to the global network connectivities in Figure 4.2 is the relative unimportance of Pacific Asia, which has only five dominant centres compared to fourteen for Western Europe and eleven for northern America.

Global command centres

The failure of Pacific Asia as a region of powerful world cities (in the original sense of 'command and control centres') is further accentuated when the focus turns to headquarters cities, those with service values of 5. There are only twenty-one cities that house the headquarters of the 100 global service firms: they can be properly termed 'command centres'. These are shown in Figure 4.8 and are differentiated in terms of the total product of service values that includes a city's scores of 5. Once again London and New York stand out, with values of 21,920 and 17,649 respectively, and the third place is far behind (Chicago with 5,145). These two cities are the 'mega' command centres of the world city network. There are two other cities that have command products above 4,000 and, with Chicago, are designated 'major' command centres. The boundary point between 'medium' and 'minor' is 1,000. The key feature of this pattern is the total concentration in Western Europe and northern America with the sole exception of Tokyo as a major command centre. On this occasion there are as many cities represented in northern America as in Western Europe (10 each). This is a stark picture of where the direct instrumental power lies within the world city network. Globalization may be a worldwide phenomenon but its command centres are most certainly not so distributed.

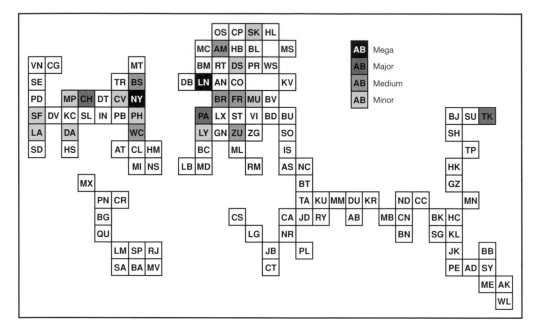

Figure 4.8 Global command centers (from Taylor *et al*. 2002c). (For city codes, see Figure 4.1.)

Regional command centres

The degree of power exercised by headquarters offices will obviously vary by firm depending upon how each firm organizes its decision-making processes (Dicken 1998). Some firms are'vertical' in their structures (a hierarchy of offices), whereas others are more 'horizontal' (relatively autonomous offices). For instance, the law firm White & Case concentrates its decision making in its New York headquarters, whereas its rival Baker & McKenzie is very decentralized and refuses to call its central 'administrative office' in Chicago the headquarters (Beaverstock *et al*. 2000b). Both hierarchy and decentralization can take a geographical form, with particular offices chosen to be 'regional headquarters'. The operative word here is 'chosen'. Generally, the main headquarters reflects a firm's origins and it is usual for the city where a firm began to continue to house its main decision-making functions. In contrast, regional headquarters are designated as part of a firm's spatial strategy and are therefore particularly relevant for searching out global patterns (Godfrey and Zhou 1999).

Most of the global service firms in the data do not have designated regional offices, but there are enough that do to show a clear pattern of spatial organization. As part of the data collection, all offices that had 'extra-locational functions' were identified; these could be national, transnational or regional in nature. Here I concentrate upon the 118 transnational and regional offices. Ignoring cities with just one such office (which just reflects particularities of a single firm), cities with transnational and regional offices are shown in Figure 4.9. London has by far the most of such offices (25), often with responsibilities covering Europe, the Middle East and Africa (EMEA). If 'major' regional world cities are defined as those with at least ten such offices, there are three other cities that qualify: Hong Kong (15½, the '½' indicating shared responsibility), New York (13) and Miami (11). There are also three cities designated 'medium' regional world cities, with from seven to nine such offices: Singapore (9), Tokyo (7½) and São Paulo (7½).

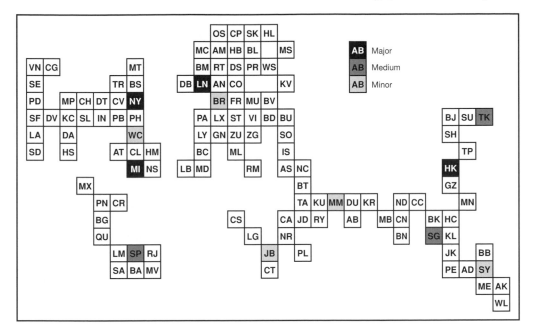

Figure 4.9 Regional command centers (from Taylor *et al.* 2002c). (For city codes, see Figure 4.1.)

These results are very similar to those given in Taylor (2000a) but are based upon much more evidence. The remaining cities in Figure 4.9 have from two to four transnational or regional offices. Notice that with this form of organization there is a diffusion of some instrumental power out from northern America/Western Europe. Pacific Asian cities reappear beyond Tokyo, and there are three cities servicing 'southern regions' (São Paulo, Johannesburg and Manama). However, there are still *extra-regional* headquarters, notably London for the Middle East/Africa and Miami for the Caribbean/Latin America, where 'Northern' power impinges directly into the 'Southern' continents even at this regional scale.

Conclusion on domination and command: from this power perspective, Sassen's notion of global cities transcending the North–South divide seems a trifle sanguine; globalization begins to look very 'Western' as soon as direct expressions of power are investigated.

Network power: gateway cities

The existence of an infrastructural power through the network is clearly suggested by the major discrepancy between the network connectivity rankings and the command functions. Whereas Hong Kong ranks third in global connectivity, it has no global command functions. This means that despite that lack, Hong Kong has attracted large numbers of service firms because of its position in the network. It is the prime location for firms to service clients in the growing Chinese market. Thus for many a global strategy, Hong Kong is a place where you *have to be*. Hong Kong is the node in the network where specialist knowledge on abilities and possibilities in the Chinese market intersects with global flows of information and ideas. Places such as this, where firms need to be to service their clients, embody a network power through their network and geographical position.

Traditionally, such 'necessary regional cities' have been called gateway cities (Johnston, 1982), and this terminology has entered the world city literature (Drennan 1992; Mayerhofer and Wolfmayr-Schnitzer 1997; Drbohlev and Sykora 1999; Andersson and Andersson 2000; Short *et al.* 2000). Quite simply, the world economy does not consist of an undifferentiated market; there are congeries of regional and national markets each with their own particularities that have to be translated through gateway cities. Gateway cities are defined in two different ways drawing upon previous analyses.

High-connectivity gateways

The simplest way to define the places where many firms decide they 'have to be' is to look beyond the twenty-one command centres (Figure 4.8). In Figure 4.10 the top thirty-five cities *without* command functions have been selected in terms of their high global network connectivity. As already noted, Hong Kong ranks number one here since it is the third most connected world city even though it has no global command functions. These highly connected non-command centres are divided into three levels in terms of their network connectivity in Figure 4.10. At the highest level the cities each relate to a major national economy outside the top five economies (the United States, Japan, Germany, France, the United Kingdom): as well as Hong Kong/China there are Milan/Italy, Toronto/Canada, Madrid/Spain, Sydney/Australia and São Paulo/Brazil plus the regional Singapore/ASEAN. Cities at the next level have a similar relation: Mexico City/Mexico, Buenos Aires/Argentina, Mumbai/India, Taipei/Taiwan plus the regional Miami/Caribbean–Latin America. These are all cities attending to gateway functions for national and regional markets: they are the classic gateway cities of contemporary globalization.

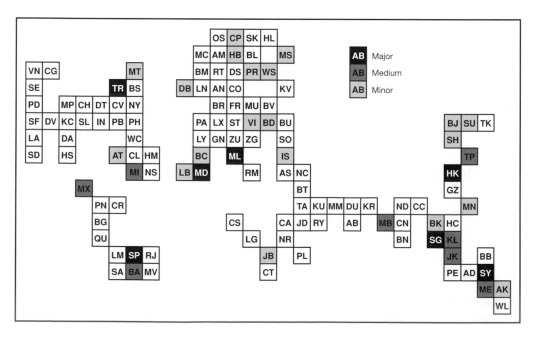

Figure 4.10 High-connectivity gateways (from Taylor *et al.* 2002c). (For city codes, see Figure 4.1.)

Gateways to emerging markets

An alternative approach is to look at connectivity-through-subordination. This is the obverse of searching for dominant cities (Figure 4.7). Based as they are upon the subordinate relations, there is a temptation to see these cities only as 'dependent' within a hierarchy. But in network relations, where all cities are dependent on all others by definition, this subordination does not equal powerlessness. Rather I interpret these cities as 'emerging centres', new strategic places where firms from elsewhere choose to expand their geographical reach. In Figure 4.11 thirty-one cities with connectivity-through-subordination levels above 5,000 are shown; those with levels above 6,000 are selected as 'major' emerging centres. By definition, all these cities have few important offices – global or regional – but they house large numbers of ordinary offices. This suggests that the cities each have a particular attraction to many global service firms that have to have a presence in the city. Beijing has the highest connectivity-through-subordination, followed closely by Moscow. These are obviously capital cities of countries with large 'emerging' markets. Other major emerging centres – Seoul, Caracas and São Paulo – are also leading cities in important emerging markets. Zurich, Europe's only major emerging centre, is a special case relating to Switzerland's success as a 'neutral' venue (especially in banking, where it is more a 'lax gatekeeper' than a gateway) within the world-economy. Beyond these major emerging centres the other cities in Figure 4.11 are quite similar in nature, being leading cities in emerging markets outside or on the fringe of the core of the world-economy.

Conclusion for network power: there are cities that are important strategic nodes within the world city network but which have no command power. In this configuration of power, the world city network does appear to transcend the North–South divide power differentials to a measurable degree.

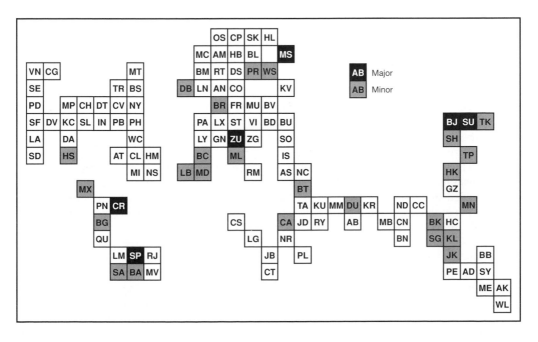

Figure 4.11 Gateways to emerging markets (from Taylor *et al.* 2002c). (For city codes, see Figure 4.1.)

Comparative connectivities

The world city network as defined by advanced producer services is not the only way in which contemporary cities are linked across the world. In this book it is argued that financial and business services are currently leading economic sectors that are dominating world city network formation, but firms in these sectors are by no means the only 'world city networkers'. The enabling technologies in computers and communications may have been stimulated in many of their advances through the demand created by financial and business services, but the technologies remain available to others. Worldwide communities relating to environmental issues and other global campaigns have created networks through non-governmental organizations (NGOs). All such institutions are instrumental in linking up their activities across the world, largely through cities. In fact, some of the larger NGOs have office networks as large and widespread as those of many global service firms; they are creating their own interlocking network of cities.

The interlocking network model can be used as a general conceptualization of city network formation involving a range of city networkers. Thus the methodology described in the previous chapter can be applied to all groups of organizations that have established networks of presence across cities. In this way it can be said that, say, environmentalist campaigners with their myriad global organizations are city network builders, just like business service firms. Thus as with the latter as described previously, data can be collected and manipulated to define new connectivities of cities reflecting a different set of inter-city relations. Here I look at three alternative ways in which cities are connected globally and compare the resulting city connectivities with global network connectivities (i.e. business service connections).

First, I consider global media cities as defined by the huge media conglomerates that have formed in the past decade. The results of feverish take-over activities, these firms now combine television and film with newspapers and books, and with web services and advertising. In the latter category there is a small overlap with the global services data. The data have been collected by Stefan Kratke (2002) and cover the leading thirty-three global media companies and their presence in 196 cities across the world. He has produced measures of the size of each firm's presence in each city, from which global media network connectivities are computed (using equation 3.10). Focusing on the top 104 cities (those with at least five firms present), a new cartogram has been produced to portray the new geography of connectivities (Figure 4.12). Compared with the original 'archipelago' of global service centres (Figure 4.1), media cities have a particularly European bias to their distribution, with both minor US cities and African/Middle Eastern cities missing from the new cartogram.

This European orientation to global media organization is confirmed by the new global urban geography in Figure 4.13. To show the pattern of media city connectivities I have divided the cities into three groups: the top twenty-five, the next twenty-five and the rest (i.e. those outside the top fifty cities measured for media connectivity). Well over half of the top group (16) are European cities. There are only three US cities in this group plus one each from Canada and Latin America, and three from Pacific Asia plus one from Australia. The explanation for this pattern is simply that although media firms might be organized globally, media markets tend to be very national in scope. Hence the global urban geography reflects a strategy for locating in the main cities of this mosaic of markets. The sixteen European cities are the leading cities of the largest national markets plus Barcelona as the media centre for Spanish Catalonia, the largest European media market not defined by state boundaries. In comparison to Europe, the US media market is more unified and therefore global media companies do

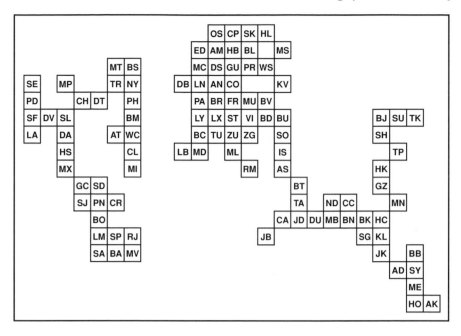

Figure 4.12 An archipelago of media cities. The cartogram places cities in their approximate relative geographical positions. The codes for cities are as follows: AD Adelaide; AK Auckland; AM Amsterdam; AN Antwerp; AS Athens; AT Atlanta; BA Buenos Aires; BB Brisbane; BC Barcelona; BD Budapest; BJ Beijing; BK Bangkok; BL Berlin; BM Baltimore; BN Bangalore; BO Bogotá; BR Brussels; BS Boston; BT Beirut; BU Bucharest; BV Bratislava; CA Cairo; CC Calcutta; CH Chicago; CL Charlotte; CO Cologne; CP Copenhagen; CR Caracas; DA Dallas; DB Dublin; DS Düsseldorf; DT Detroit; DU Dubai; DV Denver; ED Edinburgh; FR Frankfurt; GC Guatemala City; GU Gütersloh; GZ Guangzhou; HB Hamburg; HC Ho Chi Minh City; HK Hong Kong; HL Helsinki; HO Hobart; HS Houston; IS Istanbul; JB Johannesburg; JD Jeddah; JK Jakarta; KL Kuala Lumpur; KV Kiev; LA Los Angeles; LB Lisbon; LM Lima; LN London; LX Luxembourg City; LY Lyons; MB Mumbai; MC Manchester; MD Madrid; ME Melbourne; MI Miami; ML Milan; MN Manila; MP Minneapolis; MS Moscow; MT Montreal; MU Munich; MV Montevideo; MX Mexico City; ND New Delhi; NY New York; OS Oslo; PA Paris; PD Portland; PH Philadelphia; PN Panama City; PR Prague; RJ Rio de Janeiro; RM Rome; SA Santiago; SD Santo Domingo; SE Seattle; SF San Francisco; SG Singapore; SH Shanghai; SJ San José; SK Stockholm; SL St Louis; SO Sofia; SP São Paulo; ST Stuttgart; SU Seoul: SY Sydney; TA Tel Aviv; TP Taipei; TR Toronto; TU Turin; VI Vienna; WC Washington, DC; WS Warsaw; ZG Zagreb; ZU Zurich.

not have to locate in the smaller US cities. However, this dearth of major media cities in the United States may mask a general Western bias in this global industry. Although Pacific Asia presents a mosaic of national media markets similar to Europe, the region looks more like the United States than Europe in Figure 4.13. This pattern is explored further in Chapter 6. In terms of global media, it appears that Pacific Asian cities join with cities in other non-Western regions in a perpetuation of the old core–periphery pattern.

The second set of world city 'networkers' I consider are the new social movements as reflected in the organization of their NGOs. In particular, I focus on environmental, development, humanitarian and human rights NGOs as defined by *The UN Yearbook of International Organizations*. Using this source for 2001–02, seventy-four NGOs were selected for having offices in cities across at least three continents and for which good information could be obtained comparable to that for the global service firms. In this case the importance of cities for individual NGOs was scored from 0 to 4. The data

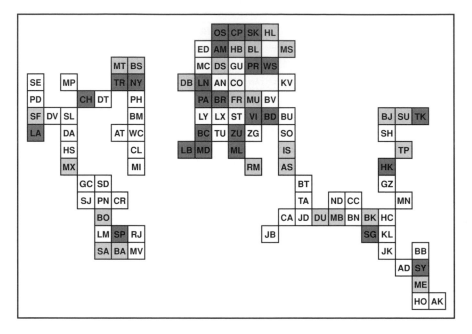

Figure 4.13 Media network connectivity (from Kratke and Taylor forthcoming). (For city codes, see Figure 4.12.)

cover more than 600 cities and towns with NGO presences. For the cartogram in Figure 4.14 I have used just the 100 cities with the highest NGO inter-city connectivity. Here is a very different distribution of cities across the world. Regions previously identified as having few or no world cities are now represented: the most obvious case is sub-Saharan Africa, which now has over a quarter of the cities on the cartogram, but notice also the appearance of Central Asia and Central America, which were both wholly missing from Figure 4.1. In contrast, the reduction of the United States to just two cities on this cartogram is quite startling; clearly, US cities are relatively very unimportant to the NGO global space of flows. The explanation for this is partly to do with the political nature of NGO business. Figure 4.14 is dominated by capital cities, locales where NGOs work with and through national politicians and governments. Thus for the United States, Washington as capital and New York as UN headquarters appear to be sufficient. Of the remaining ninety-eight cities in Figure 4.14, ninety-two are capital cities.

This second alternative global urban geography is shown in Figure 4.15 and has been constructed in the same way as Figure 4.13 to facilitate easy comparison. In this case, sub-Saharan Africa has six of the major cities, which makes it the leading region; both Western Europe and Pacific Asia have five each. However, one of the most notable features of this geography is the widespread pattern of major cities. Beyond the three leading regions, major cities are scattered across Eastern Europe (Moscow), the Middle East (Cairo), South Asia (New Delhi, Dhaka), Latin America (Mexico City, Santiago, Buenos Aires) as well as the United States (Washington and New York). If globalization is about organization that is worldwide in scope and operation, NGO connectivities show these institutions to be the globalizers *par excellence*.

For a third new geography of world city networkers, I go back to the original services data and abstract the part of the service values matrix relating to banking/finance. Using just the twenty-three banking/finance firms, inter-city connectivities are computed to

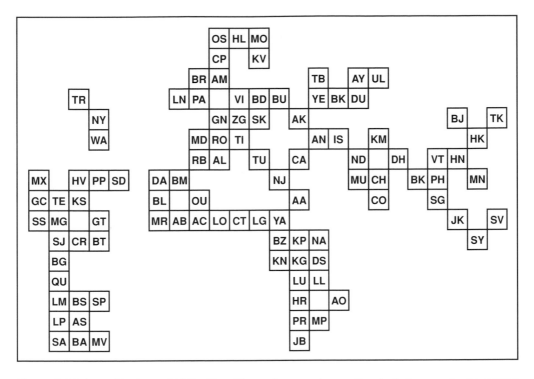

Figure 4.14 An archipelago of NGO cities. The cartogram places cities in their approximate relative geographical positions. The codes for cities are as follows: AA Addis Ababa; AB Abidjan; AC Accra; AK Ankara; AL Algiers; AM Amsterdam; AN Amman; AO Antananarivo; AS Asunción; AY Almaty; BA Buenos Aires; BD Budapest; BG Bogotá; BJ Beijing; BK Baku; BL Banjul; BM Bamako; BN Bangkok; BR Brussels; BS Brasilia; BT Bridgetown; BU Bucharest; BZ Brazzaville; CA Cairo; CH Chennai; CO Colombo; CP Copenhagen; CR Caracas; CT Cotonou; DA Dakar; DH Dhaka; DS Dar es Salaam; DU Dushanbe; GC Guatemala City; GN Geneva; GT Georgetown (Guyana); HK Hong Kong; HL Helsinki; HR Harare; HV Havana; IS Islamabad; JB Johannesburg; JK Jakarta; KG Kigali; KM Kathmandu; KN Kinshasa; KP Kampala; KS Kingston; KV Kiev; LG Lagos; LL Lilongwe; LM Lima; LN London; LO Lomé; LP La Paz; LU Lusaka; MD Madrid; MG Managua; MN Manila; MO Moscow; MP Maputo; MR Monrovia; MU Mumbai; MV Montevideo; MX Mexico City; NA Nairobi; ND New Delhi; NJ N'Djamena; NY New York; OS Oslo; OU Ouagadougou; PA Paris; PH Phnom Penh; PP Port au Prince; PR Pretoria; QU Quito; RB Rabat; RO Rome; SA Santiago; SD Santo Domingo; SG Singapore; SJ San José; SK Skopje; SP São Paulo; SS San Salvador; SV Suva; SY Sydney; TB Tbilisi; TE Tegucigalpa; TI Tirana; TK Tokyo; TR Toronto; TU Tunis; UL Ulan Bator; VI Vienna; VT Vientiane; WA Washington, DC; YA Yaoundé; YE Yerevan; ZG Zagreb.

provide measures of global financial connectivity. Based upon the initial cartogram of 123 cities (Figure 4.1), a further geography of connectivities is produced (Figure 4.16) using the same way of dividing cities as in Figures 4.13 and 4.15. The highly connected cities on this figure are the international financial centres (IFCs) of the world. This is to extract the IFC element from within the broader concept of world city dealt with previously. This singular abstraction is particularly interesting because of the large literature on IFCs that no other global service can begin to match. The establishment of most IFCs of importance preceded the rise of contemporary world cities (Reed 1981). But not all of them: the recent development of Pacific Asia as a core zone for business services has been built primarily on banking/finance. The growth and regional spread of Japanese banks plus the attraction of US and European banks to the region have

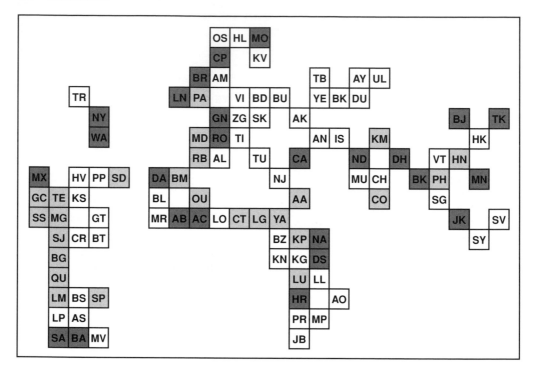

Figure 4.15 NGO network connectivity. (For city codes, see Figure 4.14.)

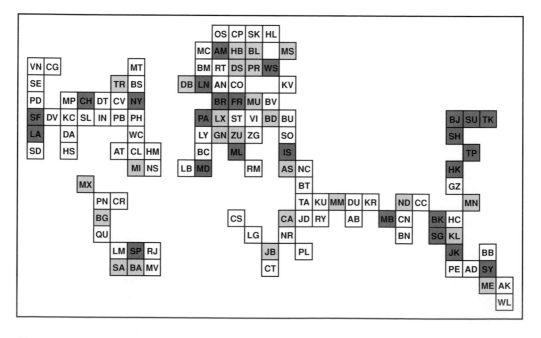

Figure 4.16 Finance network connectivity. (For city codes, see Figure 4.1.)

produced the leading world region in terms of financial global network connectivity. Of the thirteen Pacific Asian cities in Figure 4.16, no fewer than nine feature as major IFCs. No other region can match this ratio or number: Western Europe has seven and northern America just four. Thus the rise of Pacific Asian cities has been disproportionately dependent on banking/finance, as shown earlier (Figure 4.6b).

Within these general geographies there are detailed patterns that can be compared by looking at the rankings. I concentrate on the top echelons of the connectivities. In Table 4.6 the top twenty-five cities in each of the four city-interlock connectivities are listed. All lists have London and New York ranked at the top except for the NGO connectivity, where Nairobi ranks first! Surely Nairobi has never ranked this high before in a world cities study. Its pre-eminence reflects, of course, the very different nature of the global space of flows in this case. Overall, the table includes fifty-three cities, of which only four appear in all four lists: London, New York, Tokyo and Brussels. No real shock here, with leading world cities that are either important capital cities or international 'capital cities' (of the UN and EU). There are twenty-three cities that feature in two or three lists, but perhaps the interesting ones are those that appear just once. Perhaps not surprisingly, there are only two cities that are to be found only in each of the services and the financial rankings: Melbourne and Miami, which rank twenty-fourth

Table 4.6 *Rankings of cities on four network connectivities*

Rank	Global network connectivity	Bank network connectivity	Media network connectivity	NGO network connectivity
1	London	London	London	Nairobi
2	New York	New York	New York	Brussels
3	Hong Kong	Tokyo	Paris	Bangkok
4	Paris	Hong Kong	Los Angeles	London
5	Tokyo	Singapore	Milan	New Delhi
6	Singapore	Paris	Madrid	Manila
7	Chicago	Frankfurt	Amsterdam	Washington, DC
8	Milan	Madrid	Toronto	Harare
9	Los Angeles	Jakarta	Stockholm	Geneva
10	Toronto	Chicago	Copenhagen	Moscow
11	Madrid	Milan	Sydney	New York
12	Amsterdam	Sydney	Singapore	Mexico City
13	Sydney	Los Angeles	Barcelona	Jakarta
14	Frankfurt	Mumbai	Zurich	Tokyo
15	Brussels	San Francisco	Vienna	Accra
16	São Paulo	São Paulo	Oslo	Cairo
17	San Francisco	Taipei	Prague	Dhaka
18	Mexico City	Shanghai	Tokyo	Rome
19	Zurich	Brussels	Brussels	Dakar
20	Taipei	Seoul	Hong Kong	Santiago
21	Mumbai	Istanbul	Budapest	Abidjan
22	Jakarta	Beijing	Warsaw	Buenos Aires
23	Buenos Aires	Bangkok	Lisbon	Dar es Salaam
24	Melbourne	Amsterdam	Chicago	Copenhagen
25	Miami	Warsaw	São Paulo	Beijing

and twenty-fifth for global network connectivity, and Seoul and Istanbul, which rank twentieth and twenty-first for bank network connectivity. From the previous discussion of Figure 4.13 it will be no surprise that six of the seven cities that are ranked only under media network connectivity are European capital cities (the exception is again Barcelona, capital of Catalonia with its own language and media market). Not just for its number one ranking city, the most distinctive list is that for NGO network connectivity, where most (15) cities are not found in other lists. This confirms the evidence of Figure 4.15 that these inter-city relations define a quite different world city network. And it is not just the periphery bias especially featuring Africa (which has seven of the fifteen cities unique to this list) that is noteworthy. Within the United States (with Washington, DC) and Western Europe (with Geneva and Rome) there are different nodes within the core to which this network connects.

What these results show is that while the important cities in terms of global network connectivity and financial network connectivities are sometimes also important in media and NGO spheres, there are important differences with the latter networks. Clearly, cities in globalization involves more than financial and business services, but the latter are the dominant networkers and I continue to focus on them in the chapters that follow.

⑤ City network analyses

> To understand the evolution of the contemporary ways of the world, networks
> of cities are fundamental.
>
> (Gottmann 1984: 1)

Measures of connectivity are the first clear benefit to be obtained through precise specification and customized data collection. Such measures provide a basic understanding
of the cities within a network; they go a long way to satisfying the evidential lacuna in
inter-city relations identified in Chapter 2. This is where I began Part II as discussions
of 'connections'. But connectivities do not exhaust this discussion. As aggregate
measures they are good for general assessment of cities, and disaggregations can inform
us about roles and positions of cities within the network, as the previous chapter demon-
strates. What is missing is any notion of how the network fits together. Thus the previous
chapter can be viewed as 'node orientated'; in this chapter I move on to become much
more 'network orientated': inter-city relations are analysed as a network structure.

The network analyses presented in this chapter are organized at three levels of focus.
First, a particular egocentric analysis is developed that focuses upon one node at a time
and describes its specific position in relation to all other nodes. Thus by going beyond
aggregate measures of connectivity, the pattern of a city's linkages with other cities is
constructed. This provides answers to the following types of question. Does London
have more intensive connections with other European cities or with US cities? Where
are Sydney's strongest connections: to Pacific Asian cities, to US cities or to West
European cities? Hence I return again to Jane Jacobs' (1984) conception of cities having
direct influences beyond their city regions. Second, I use standard clique analysis to
identify intense 'sub-networks' within the overall network. The sorts of questions that
are answered here are at the regional scale. For instance, do the globalization arenas –
northern America, Western Europe and Pacific Asia – constitute cliques of densely
connected cities? Third, I look at the network as a whole and construct a new global
space of inter-city relations. This produces what I call a new 'landscape' of cities that
can be used to explore the relative positions of cities within the overall network. The
chapter is divided into three sections based on these levels and concludes with a brief
discussion of the problem of visualizing the world city network and the need for an
alternative medium to the book format that I am restricted to here.

Egocentric analyses: city hinterworlds

There is no readily available term for describing the pattern of a city's connections
across the world and therefore I have had to invent one: hinterworld. Defining hinterworlds in relation to other concepts describing 'urban influences' is the task of the first

sub-section below. The second sub-section considers how to measure hinterworlds and a method is devised that allows comparisons to be made between cities in terms of their patterns of connections within the world city network. The third sub-section then illustrates this comparative use of hinterworlds to show contrasts between cities in terms of globality and specificity.

Inventing a new concept

In studying the external relations of cities and towns, urban geographers and rural sociologists have traditionally considered urban places as local service centres. Although different from our global service centres in the nature of their clientele – dispensing retail services rather than producer services – the idea of an urban concentration of services for outside customers still holds. But the nature of the geography of this urban influence fundamentally changes in a global space of flows, necessitating the invention of a new concept for the new circumstance. But not too new – I will begin by considering this original geography.

Delimiting the region around a town or city for which it provides services was a major research topic in the urban geography of the 1950s and 1960s. The region was variously called sphere or zone of influence, urban field, tributary or catchment area, and umland or hinterland, but I follow Johnson (1967) and use the last of these terms. Information for defining hinterlands could be obtained either from the centre (e.g. mapping the circulation of a city newspaper) or from outside (e.g. mapping the shopping trips of rural residents). Although the purpose was typically to determine discrete regions by drawing boundaries between urban centres, in practice two general findings were commonplace: first, hinterlands were found to overlap (Johnson 1967: 87); and second, within hinterlands there were variations in intensity of influence, typically a distance decay gradient from the centre (ibid.: 89). Both overlap and varying intensity are accentuated in relation to world cities within globalization.

In world city network formation there are no boundaries between cities in terms of their hinterlands. With electronic communication underpinning their office networks (for law examples, see Beaverstock *et al.* 1999b, 2000b), every city's hinterland overlaps with every other city's hinterland. For instance, a German firm doing business in Australia can work through the 'local' office of a global bank in Hamburg, just as an Australian firm doing business in Germany can work through the 'local' office of a global bank in Adelaide. This is what economic globalization is all about and it is the reason the notion of hinterland is problematized for world city activities: IT has created worldwide overlap of urban influences. With no boundaries enclosing the clients serviced through a city, all that is left by way of geography is the varying intensity of service connections. Whether this declines by distance from a city is less likely, given IT, but this is left as an empirical question to be addressed later.

The term 'hinterworld' has been coined to reflect the geographical scale and nature of service provision by world cities. I am now in a position to define these hinterworlds: a city's hinterworld is the global distribution of service connections that lies behind its world city formation. Specifically, I will measure and map it as the spatial variation in levels of connection that one city has with each other city in the network.

Before I begin empirically investigating these hinterworlds it is important to understand that hinterworlds in no way supplant or supersede long-standing hinterlands. Hinterlands have not disappeared and will not disappear just because globalization adds new processes to the economic mix. Global retailing corporations may restructure local shopping patterns but they still operate by servicing local hinterlands. And, of course, residents in southern New Hampshire still look to Boston for their regional newspapers and sports teams irrespective of Boston having developed an important hinterworld over the past two decades.

Measuring hinterworlds

To compute a city's hinterworld it is necessary to specify its external relations with other cities. For this exercise I deal with just the top 123 cities in terms of global network connectivity. To find a satisfactory measure of hinterworlds takes two steps. I begin by describing an initial 'absolute' measure that is shown to have limited utility for comparing city hinterworlds. However, it forms the basis for a 'relative' measure that does turn out to be interesting in this respect and therefore is used for subsequent comparative hinterworld studies.

From the starting point that a city's hinterworld consists of the levels of service it provides for doing business in each of the other 122 cities, it is initially computed as follows. First, count the number of firms present in each city. For each city multiply this number by 5, the maximum service value. This constitutes the highest possible level of service that a city could expect in another city (i.e. the other city houses the head-quarters of every single one of the original city's global service firms). Thus in the simple data set shown in Table 5.1, highest levels of possible service are 20 for London (with 4 firms), 15 for Munich and 10 for Lagos. Now for each city, take other cities in turn and sum their service scores, but *only* for firms present in the original city. For instance, starting with London, the sums for Munich and Lagos are 10 and 4 respectively; starting with Munich the sums for London and Lagos are 12 and 2; and starting with Lagos the sums for London and Munich are 10 and 2. The latter sums are expressed as proportions of the highest level of possible service. For instance, the proportions for London are Munich 0.5 (= 10/20) and Lagos 0.2 (= 4/20). All such computations are shown in Table 5.2.

The interpretation of these proportions is relatively simple. The columns in Table 5.2 define the level of service that can be expected in a city when visiting a global service firm in a row city. Thus for someone going into an office in London to do business in Munich the service level is 0.5, but for someone doing business in Lagos the level falls to 0.2. In contrast, for someone starting in Munich to do business in London, the latter offers a 0.8 level of service, and in the case of Lagos, doing business in London has a 1.0 service level, showing that Lagos's two service firms in Table 5.1 both have their

Table 5.1 *A simple matrix of service values*

	Firm A	Firm B	Firm C	Firm D
London	5	5	5	2
Munich	2	3	0	5
Lagos	2	0	2	0

Table 5.2 *Levels of servicing derived from Table 5.1*

	London	Munich	Lagos
London	—	0.8	1.0
Munich	0.5	—	0.2
Lagos	0.2	0.13	—

headquarters in London. In other words, London services Munich and Lagos much better than these cities service London. Not surprisingly, the lowest level of service in this data is a paltry 0.13 for doing business from Munich in Lagos. These columns represent the absolute servicing linkages that form the basis of a city's hinterworld.

There is a basic problem when comparing the hinterworlds of cities using absolute measures and this can be easily seen in Table 5.2. Notice that in this table London appears with very high service levels for the other two cities and Lagos provides low levels. This is obviously reflecting the network position of these cities in this small data set. If this result is extrapolated to the 100 × 123 data set, it is found that every city has its highest external provision in either London or New York. In general, it can be noted that absolute hinterland measures across cities tend to follow closely the level of the global network connectivities of cities. This means that mapping absolute hinterlands produces results that largely replicate the global network connectivity distribution: all hinterworlds, although not exactly the same, look quite similar. Hence this measure is of relatively little utility when comparing cities. Small differences are discernible, but intense scrutiny of absolute hinterland maps is not the sensible way forward here. To overcome this comparative deficiency I add an extra step to describing hinterworlds.

What is required is a measure of hinterworlds from which the general network connectivity of cities is removed. Devising one is actually a relatively simple task. Scatter diagrams of absolute hinterworld scores across other cities against those cities' global network connectivities show strong positive linear relationships in every case: in other words, the higher a city's overall connectivity, the higher the external provisioning of that city to other cities. Thus a given city's absolute hinterland values across the other 122 cities can be regressed against the global network connectivity of those 122 cities using the simple regression equation

$$y_i = a + bx_i (+ R_i) \qquad (5.1)$$

where y_i is the absolute provision values for city Y across 122 other cities, x_i is the global network connectivity of the 122 other cities and R_i represents the residual. The residuals are departures from the regression line computed as the difference between the city's absolute level of hinterworld (y_i) and its predicted level (y_i') using equation 5.1 (outside the bracket) with calibrated values for the constants a and b:

$$R_i = y_i - y_i' \qquad (5.2)$$

This works as follows. Consider the hinterworld of a city P whose absolute provision value for city Q (y_q) is 0.5. City Q has a global network connectivity score of 0.2. Feeding $x = 0.2$ into the regression equation produces a predicted level of link (y_q') of 0.4 for city P's level of connectivity. The residual R_q is therefore $y_q - y_q'$ or 0.1. This is more than expected; it shows that city P is 'over-linked' to city Q relative to the latter's global network connectivity. On the other hand, if y_q had been computed as 0.6, then at $R_q = -0.1$, city P would have been shown to be 'under-linked' to city Q relative to the latter's global network connectivity. These residuals define a 'relative' hinterworld – where a city is strongly serviced and where it is weakly serviced with respect to the servicing city's position in the world city network (i.e. its connectivity). These relative measures of hinterworlds are employed in all subsequent discussion.

Real examples will illustrate the comparative potential of the new measure. Results from such a residual analysis for London and Manchester are shown in Figures 5.1–5.4. The scatter diagrams (Figures 5.1 and 5.2) show that for London, the largest positive residual is Washington, DC, and for Manchester it is Birmingham. This suggests a contrast between the London's 'globality' and Manchester's more intense 'local'

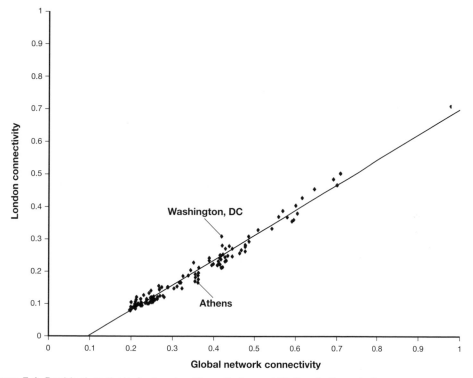

Figure 5.1 Residual analysis for London (from Taylor and Walker forthcoming)

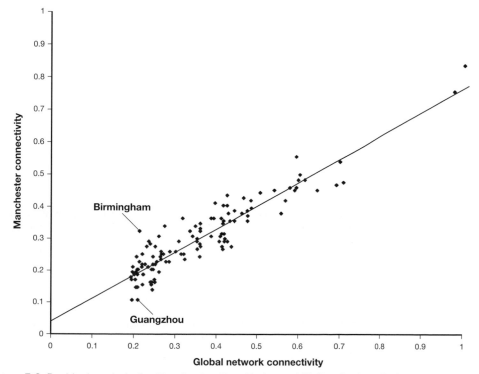

Figure 5.2 Residual analysis for Manchester (from Taylor and Walker forthcoming)

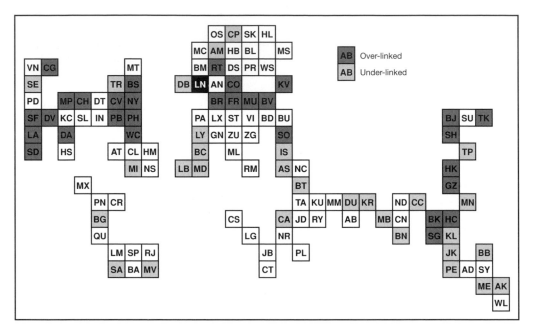

Figure 5.3 London's hinterworld (from Taylor and Walker forthcoming). (For city codes, see Figure 4.1.)

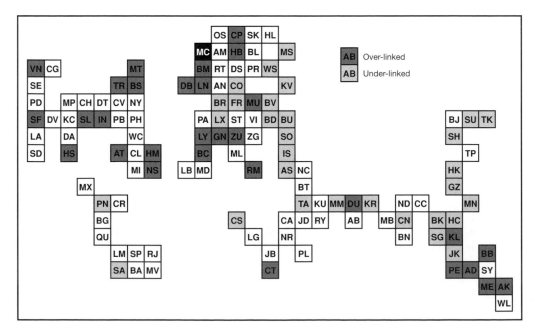

Figure 5.4 Manchester's hinterworld (from Taylor and Walker forthcoming). (For city codes, see Figure 4.1.)

relations. This idea is augmented by their respective largest negative residuals, with London's being in Europe (Athens) and Manchester's being in China (Guangzhou). In Figures 5.3 and 5.4 the two hinterlands are mapped by showing the thirty cities with the largest positive residuals as over-linked and those thirty with the largest negative residuals as under-linked. These maps of 'relative' hinterworlds (Figures 5.3 and 5.4) show that the global versus local interpretation of London's and Manchester's connections is only partly confirmed. To be sure, London shows itself to be very strongly linked to US and Pacific Asian cities and certain European cities (particularly German ones). Manchester, on the other hand, also shows a worldwide pattern of strong linkages, but this time it reflects historical British ('Old') Commonwealth connections, with Canadian and Australian cities very well represented. In the case of London these historical links are missing, overwhelmed by more recent developments in London's global connections, especially in banking. The latter is especially reflected in the Pacific Asian over-linkages, which is exactly where Manchester is under-linked. Both cities have strengths in the United States and Europe but with clear differences: once again, London's greater connections with German cities reflect its global banking prowess compared to that of Manchester.

The hinterworlds of London and Manchester have been used for illustrative purposes but this method can be employed substantively to compare patterns of city connections for all 123 cities.

Comparative hinterworlds

Regression analyses produce more than just residuals. If we compare the two scatter diagrams for London and Manchester (Figures 5.1 and 5.2), one feature stands out: there is a much greater scatter of points around Manchester's trend line than on the London graph. Standard errors of estimate measure the scatter about a regression line. They indicate the degree to which a city's hinterworld diverges from the global network connectivity pattern. The values for London and Manchester are 0.017 and 0.043 respectively, thus giving a precise measure of Manchester's greater spread: it is more than twice as scattered as London's graph. This can be interpreted as an indication of the degree to which specific factors tend to make a city's hinterworld diverge from the general pattern of global network connectivity. In this case there are more specific factors affecting Manchester's hinterworld than London's hinterworld. As the previous discussion has shown, Manchester's primary over-linkage is to Old Commonwealth cities; this is just the type of specific effect that will cause an increased scatter of points around a regression line.

For the whole set of 123 cities the standard errors vary between 0.013 (Madrid) and 0.067 (Indianapolis). These measures can be interpreted as a globality–specificity scale. Low standard errors show that a city's hinterworld across other cities closely matches their respective levels of connectivity. Thus it can be argued that because London's hinterworld closely reflects overall global network connectivity, it is highly globalized. In Figure 5.3 most highly connected cities are found to be over-linked: New York, Hong Kong, Tokyo, Singapore, Chicago, Los Angeles, Frankfurt, Brussels and San Francisco are all featured. In contrast, a high standard error indicates that there are many exceptions to the simple global trend reflecting specific non-global influences on a city's hinterworld – for example, Manchester's Commonwealth links. Thus in the latter case there are few highly connected cities found to be over-linked, London being the main exception (Figure 5.4). In Figure 5.5 the standard errors of all 123 cities are portrayed in three equal categories. As expected, the low standard errors representing global hinterworlds include all the most connected cities: the highest-connecting cities not in this category are San Francisco and Taipei, ranked seventeenth and twentieth respectively. The opposite obtains for high standard errors: this category is dominated by less connected cities whose hinterworlds are thus more specific in their patterns. I have chosen a further eight hinterworlds to illustrate further the globality–specificity scale.

I will start with New York (standard error = 0.018) and Tokyo (standard error = 0.017) because these cities figure prominently in the world cities literature alongside London after Sassen's (1991) identification of them as global cities. The first point to make is that these two new hinterworlds (Figures 5.6 and 5.7) are very similar to London's (Figure 5.3), with a common pattern of over-linkage that is strongest in Pacific Asia and is selective in the United States and Western Europe. Each of the three cities is over-linked to each other and they are all over-linked to Hong Kong, Singapore, Chicago, Los Angeles, Frankfurt and Washington, DC. There are no strong regional patterns to the under-linked cities and the only common feature is the under-linkage to South European cities. Overall, however, the three hinterlands can be described as core-orientated patterns, the only exception being Tokyo's over-linkage to Kuwait and Manama.

More specific hinterworlds are found to be quite regional in nature, so Manchester's more worldwide pattern is not typical. Indianapolis has the most specific hinterworld and this is illustrated in Figure 5.8. All other twenty-two US cities feature as over-linked, and beyond this home region, none of the non-US, highly connected cities make up the other eight over-linked cases. Indianapolis is under-linked to Pacific Asian cities, indicating weak banking connectivity, and East European cities, indicating weak connections to post-Soviet economic transformations. This regional localism is not just a US phenomenon: in Figure 5.9 the hinterworld of Cologne (standard error = 0.049) shows the equivalent pattern for a less connected West European city. Note that the cities of the British Isles do not feature as over-linked, with Manchester actually recorded as the only European city that is under-linked. As with Indianapolis, Cologne's connections to Pacific Asia are weak (with just two exceptions), indicating the city's low prowess in banking.

Most other US cities have hinterworlds similar to that of Indianapolis, although two cities show particular hinterwords reaching out to other parts of the world: Miami (standard error = 0.020) and Los Angeles (standard error = 0.021). Miami's hinterworld (Figure 5.10) has a clear Latin American focus, leading to rather less of a concentration on other US cities. For Los Angeles (Figure 5.11) there is over-linkage to specific Pacific Asian cities (plus Sydney) but with a stronger US city pattern than Miami. Both cities have a concentration of under-linked cities in Europe. Neither of these hinterworld patterns can be considered surprising – that is to say, they do not constitute new findings, but they can be used to support the validity of results from this hinterworld method of analysis. In addition, these two US city outreaches have never previously been illustrated so explicitly within a global urban analysis.

Finally, I will illustrate two 'Third World' cities. Manama (standard error = 0.038) and Mumbai (standard error = 0.019) are both cities for which we know much less about their global connections, and therefore hinterworld analysis should provide new findings in these cases. Manama has been promoted by Bahrain as the international financial centre for the Middle East and this is reflected in the specificities of its hinterworld (Figure 5.12). It is over-linked to all Pacific Asian cities outside China and to specific other financial centres including New York, Zurich, Geneva, Luxembourg and Panama City. Interestingly, it is under-linked to London. Overall, the main group of under-linked cities is to be found in the United States, but with the important exception of Houston for its oil interests. Whereas Manama is a specialist 'Third World' city, Mumbai is one of the few major 'Third World' cities. It is over-linked to cities in Pacific Asia but not

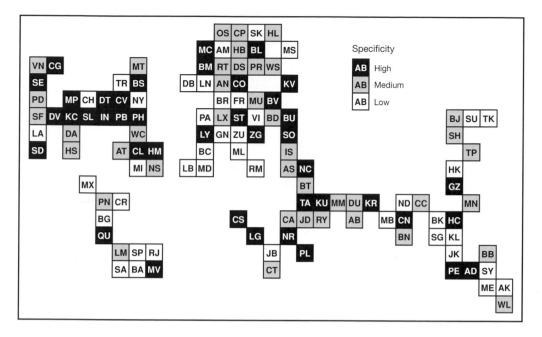

Figure 5.5 Hinterworlds: a globality-specificity scale. (For city codes, see Figure 4.1.)

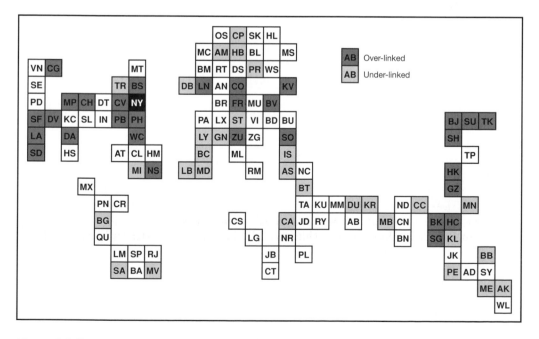

Figure 5.6 New York's hinterworld. (For city codes, see Figure 4.1.)

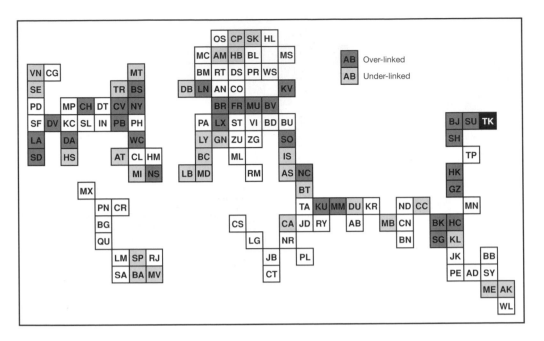

Figure 5.7 Tokyo's hinterworld. (For city codes, see Figure 4.1.)

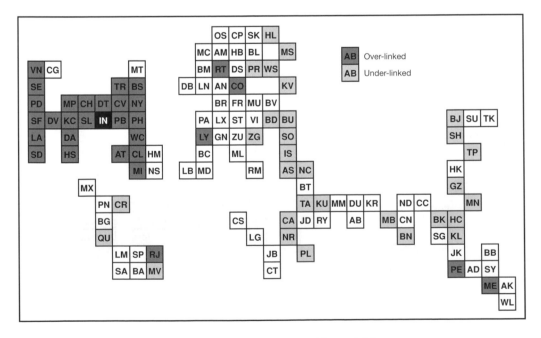

Figure 5.8 Indianapolis's hinterworld. (For city codes, see Figure 4.1.)

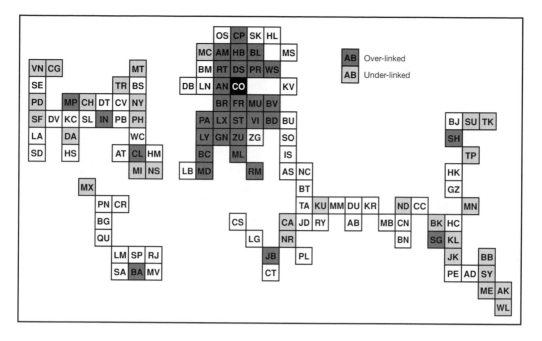

Figure 5.9 Cologne's hinterworld. (For city codes, see Figure 4.1.)

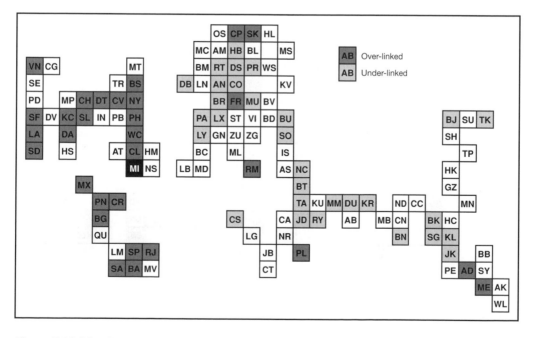

Figure 5.10 Miami's hinterworld. (For city codes, see Figure 4.1.)

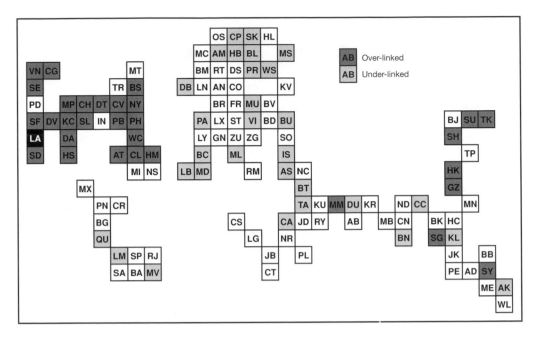

Figure 5.11 Los Angeles's hinterworld. (For city codes, see Figure 4.1.)

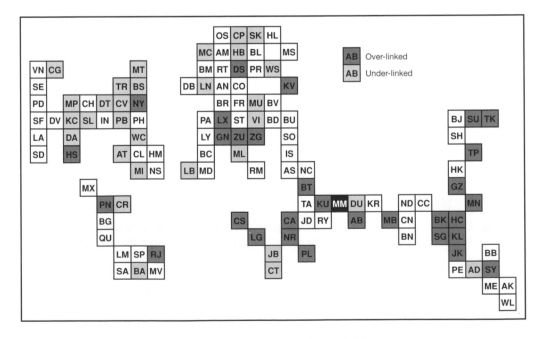

Figure 5.12 Manama's hinterworld. (For city codes, see Figure 4.1.)

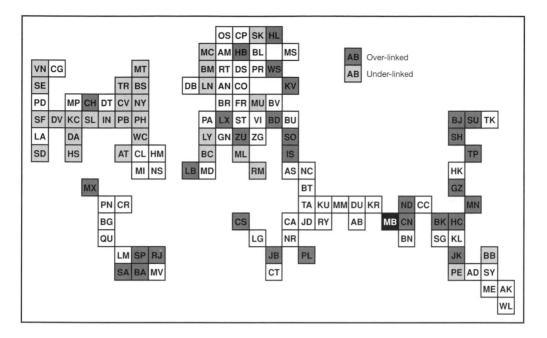

Figure 5.13 Mumbai's hinterworld. (For city codes, see Figure 4.1.)

including the big three: Tokyo, Hong Kong and Singapore (Figure 5.13). Elsewhere it is over-linked to other 'Third World' cities and to East European cities. Mumbai has an extremely weakly linked relation with most US cities (including New York but excepting Chicago) and also with British cities (including London). This hinterworld is particularly interesting because it depicts a global hinterworld quite unlike that of London, New York and Tokyo: Mumbai has a very non-core-like orientation to its pattern.

The above discussion has included just ten city hinterworlds out of 123 possible cities. All hinterworlds are unique because they depend upon the particular mix of global service firms in a city, which in turn will reflect historical and regional processes as well as contemporary location in the world city network. Thus for readers with a particular interest in one of the world cities not illustrated here, this chapter may be assumed to be quite frustrating. Providing 123 city hinterworld figures within the text is unthinkable; to locate them in an appendix seems more sensible but is also quite extravagant in space for a book that does not aspire to be an atlas of any sort. The solution is to illustrate all hinterworlds in a place where space is not a problem: in a virtual atlas on the GaWC web site www.lboro.ac.uk/gawc. Here a set of more detailed hinterworlds can be found for all 123 cities. The virtual atlas enables users to make any city comparisons, whether of 'local rivals' such as Houston/Dallas or Istanbul/Athens, or wider comparisons such as between the smaller international financial centres, say Panama City/Luxembourg City. As far as this chapter is concerned, this offer of further egocentric analyses and comparisons represents the end of the egocentric focus.

City cliques

One of the major benefits of precise specification of the inter-city network is that it allows inter-city matrices to be constructed (Taylor 2001a). In effect, this means that the interlocking network can be analysed just like any other, more conventional network. Identification of 'cliques' within connection matrices is a standard tool of network analysis and has been applied to airline connections between world cities (Shin and Timberlake 2000). Here I apply clique analysis to new inter-city matrices based upon service connectivities. In this way I take the analysis beyond aggregate connectivities and egocentric patterns.

This section is divided into two. First, I show how an inter-city connection matrix is constructed from a service values matrix and apply this to data on the 123 cities. The main features of the resulting 123 × 123 inter-city matrix are described. Second, I report on the results of applying clique analyses to this inter-city matrix. This technique searches out global cliques of world cities within the overall network. The initial results are quite stark: London and New York appear in all global cliques. This reflects the dominance of these two cities within the network; to explore beyond this finding, city cliquishness is defined and measured for all other cities.

Constructing a connection matrix for an interlocking network

This subsection is a continuation of the specification begun in Chapter 3. As before, I will use the simple example of three firms in ten cities to help explicate the argument (Table 3.1 and Figure 3.2). The example provides concrete results, but it is necessary to stress again that these are not findings; they should be treated as strictly illustrative. Meaningful findings are delayed until the next subsection.

Chapter 3 set up the specification and took the argument through to defining network connectivity. But this does not take full advantage of all the dyadic information available from equation 3.5. I repeat the argument leading to this equation for pedagogic convenience. To recap, the service value between city i for firm j is denoted as v_{ij} and the array of all such values defines the service value matrix, $\mathbf{V}$, defined by n cities and m firms. From this matrix the basic relational element for each pair of cities, a and b, is derived as

$$r_{ab,j} = v_{aj} \cdot v_{bj} \qquad\qquad (3.4/5.3)$$

This is the relation between cities a and b in terms of firm j, which I called the *elemental interlock* between the two cities for one firm. The aggregate *city interlock* is then produced from

$$r_{ab} = \sum_j r_{ab,j} \qquad\qquad (3.5/5.4)$$

This means that for every pair of cities there is a measure of connection based upon their respective service mixes. From these measures a square ($n \times n$) relational matrix, $\mathbf{E}$, is produced using the aggregate city interlock link, r_{ab} (equation 3.5/5.4), for every pair of cities. This is the *elemental relational matrix*. It indicates the quantity of intra-firm global service links between cities.

For complete matrix specification it is necessary also to define the 'self-relation' of each city to define the matrix diagonal. This is not derived from elemental interlocks as given in equation 3.4/5.3. For dimensional equivalence these self-relations are defined as

$$r_a = \sum_j v_{ja}^2 \qquad\qquad (5.5)$$

which is simply the sum of the squares of the service values in a city.

The relational matrix is illustrated by Table 5.3, where it can be seen that the highest sum is for a self-relation (New York), with London–New York showing the highest inter-city link of 20. In contrast, the link between Frankfurt and Singapore is only 3.

Table 5.3 *Elemental relational matrix*

Cities:	1	2	3	4	5	6	7	8	9	10
Cities:										
1	13	0	0	15	15	2	15	2	3	5
2	0	9	3	3	0	3	6	0	3	3
3	0	3	1	1	0	1	2	0	1	1
4	15	3	1	19	18	4	20	3	4	7
5	15	0	0	18	18	3	18	3	3	6
6	2	3	1	4	3	2	5	1	1	2
7	15	6	2	20	18	5	22	3	5	8
8	2	0	0	3	3	1	3	1	0	1
9	3	3	1	4	3	1	5	0	2	2
10	5	3	1	7	6	2	8	1	2	3

Key: 1 Chicago; 2 Frankfurt; 3 Hong Kong; 4 London; 5 Los Angeles; 6 Milan; 7 New York; 8 Paris; 9 Singapore; 10 Tokyo

This is easily interpretable as the quantity of service that a client can expect in either city for servicing business with the other city. Clearly, in this example walking into a service business firm in London to do business in New York is likely to lead to a much higher quantity of service available than going into a Frankfurt global service firm office to do business in Singapore. Note that there are some zeros in this illustrative matrix (e.g. Chicago–Hong Kong), indicating two cities sharing no firms, but this feature does not appear in the analyses of the actual service values matrix that follows.

These sums are more interpretable if transformed into a *proportionate relational matrix*, **Q**, where linkages are given as proportions of the largest linkage in the matrix. This produces a new array of $n \times n$ *proportional city interlinks* in a range between 0 and 1. This matrix is illustrated in Table 5.4, wherein the two inter-city links referred to above are converted to 0.91 (London–New York) and 0.14 (Frankfurt–Singapore);

Table 5.4 *Proportional relations matrix*

Cities:	1	2	3	4	5	6	7	8	9	10
Cities:										
1	0.48	0.00	0.00	0.56	0.56	0.07	0.56	0.07	0.11	0.19
2	0.00	0.33	0.11	0.11	0.00	0.11	0.22	0.00	0.11	0.11
3	0.00	0.11	0.03	0.03	0.00	0.03	0.07	0.00	0.03	0.03
4	0.56	0.11	0.03	0.70	0.67	0.15	0.74	0.11	0.15	0.26
5	0.56	0.00	0.00	0.67	0.67	0.11	0.67	0.11	0.11	0.22
6	0.07	0.11	0.03	0.15	0.11	0.07	0.19	0.03	0.03	0.07
7	0.56	0.22	0.07	0.74	0.67	0.19	0.81	0.11	0.19	0.30
8	0.07	0.00	0.00	0.11	0.11	0.03	0.11	0.03	0.00	0.03
9	0.11	0.11	0.03	0.15	0.11	0.13	0.19	0.00	0.07	0.07
10	0.19	0.11	0.03	0.26	0.22	0.07	0.30	0.03	0.07	0.11

Key: 1 Chicago; 2 Frankfurt; 3 Hong Kong; 4 London; 5 Los Angeles; 6 Milan; 7 New York; 8 Paris; 9 Singapore; 10 Tokyo the very high proportion for the first link is compared now with a little over one-tenth for the second link. This is the form of the matrix that is used in subsequent analyses of the service value matrix defined by the 123 cities.

To get an initial glimpse of this 123 × 123 inter-city matrix I have abstracted all proportionate links above 0.5 in Table 5.5. Of course, the London–New York link has the highest aggregate inter-city link and there are only twenty-seven other links with values half or more than half this link. These are totally dominated by links to the two leading cities: the first link that includes neither London nor New York – that between Hong Kong and Tokyo – ranks only twenty-fourth. 'Third World' cities appear only once, right at the bottom of the table, with the London–São Paulo link scoring exactly 0.5. Obviously this connection matrix is highly concentrated on the two leading cites and, beyond them, the focus is on cities in the core of the world-economy. These features will be brought out clearly in clique analyses.

Table 5.5 *Proportional inter-city links above 0.5*

Rank	Inter-city link	Score
1	London–New York	1.000
2	London–Hong Kong	0.737
3	New York–Hong Kong	0.707
4=	London–Paris	0.691
4=	New York–Tokyo	0.691
6	London–Tokyo	0.690
7	London–Singapore	0.667
8	New York–Paris	0.647
9	New York–Singapore	0.639
10	London–Chicago	0.601
11	New York–Chicago	0.600
12	New York–Los Angeles	0.593
13	London–Los Angeles	0.574
14	London–Frankfurt	0.570
15	London–Milan	0.568
16	London–Brussels	0.542
17=	London–Sydney	0.541
17=	New York–Frankfurt	0.541
19	London–Amsterdam	0.540
20	New York–Sydney	0.538
21	New York–Milan	0.537
22	London–Madrid	0.536
23	London–Toronto	0.533
24=	New York–Toronto	0.522
24=	Hong Kong–Tokyo	0.522
26=	New York–Madrid	0.515
26=	New York–Brussels	0.515
28	London–São Paulo	0.500

Cliquishness in the world city network

In network analysis a clique is defined as a set of nodes in which each is connected to all the others (Scott 1991: 117). It is normally depicted on a diagram as a 'complete graph', a set of nodes all connected to each other. In Figure 5.14 the network includes two complete graphs, A and B, each of which constitutes a clique and, in addition, B as a four-point complete graph automatically encompasses two three-point complete graphs. Note that the graph depiction assumes that all links are based upon binary data: nodes are either connected or not connected. However, the links in the inter-city matrix derived above are variable and define a 'valued graph'. Thus in order to carry out a clique analysis a valued graph has to be converted into a binary graph: the values of links have to be dichotomized so that only values above a selected threshold constitute connections (ibid.: 113). Once this conversion is made, clique analysis can proceed.

In Table 5.5 all links with values over 0.5 are listed; this is equivalent to making 0.5 the threshold so that the connections in the table define a binary graph (Figure 5.15). Cliques can be easily discerned from this graph by inspection. There is one four-city clique – London, New York, Hong Kong, Tokyo – encompassing two three-city cliques, and ten separate three-city cliques consisting of London and New York plus one of each of Paris, Singapore, Chicago, Los Angeles, Frankfurt, Milan, Brussels, Sydney, Toronto and Madrid in turn. Both Amsterdam and São Paulo record links in Table 5.5 and Figure 5.15, but only to London and therefore they do not feature in cliques. These thirteen cliques are just the tip of the iceberg: by setting the threshold at 0.5 the resulting twenty-eight connections represent a minuscule proportion of the total of 7,503 connections in the proportionate relational matrix for the 123 cities. Obviously, to increase connections for analysis the binary threshold needs lowering and this will inevitably create more cliques.

The drawback to using more links to find more cliques is that this also leads to many overlaps between cliques. Even in the preliminary clique analysis above, all identified cliques overlap through common membership of London and New York. The problem is that with more links, clique analysis can generate 'long lists of overlapping cliques,

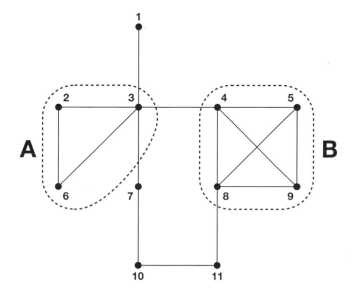

Figure 5.14 Cliques as complete graphs

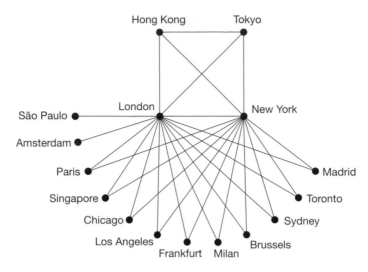

Figure 5.15 Global cliques of cities with a binary threshold of 0.5 (from Derudder and Taylor 2003)

and these results may be difficult to interpret' (Scott 1991: 122). This effect is shown in Table 5.6, in which the binary threshold is progressively reduced to reach a total of 117 clusters with a threshold of 0.15. The high degree of overlapping of clusters is shown by how many different cliques each of the top five world cities belong to. Notice, first of all, that London and New York belong to all cliques: their immensely high levels of connectivity in the global urban analysis mean that it is not possible to create a clique in which they are not both members. In this sense it can be said that 'London and New York are everywhere in the world city network'. The interesting results from a clique analysis are, therefore, to be found in memberships of other cities. For instance, in Table 5.6 it can be seen that with a threshold of 0.3, Hong Kong and Tokyo are members of nearly twice as many cliques as Paris. This is an interesting finding; it shows that Paris is particularly weak in strong connections (i.e. over 0.3) relative to the two Pacific Asian cities. Thus Paris is less cliquish in the world city network at this level of linkage.

Paris's relatively low cliquishness is removed as the binary threshold is decreased (Table 5.6), but the concept of cliquishness remains a useful one. In Table 5.7 clique memberships of the top twenty world cities are shown with a binary threshold of 0.2. Notice that the number of cliques declines broadly with connectivity rank: the higher the global network connectivity of a city, the more cliques it belongs to. But the

Table 5.6 *Clique membership of leading cities for different binary thresholds*

Binary threshold	Number of cliques	Clique membership				
		London	New York	Hong Kong	Paris	Tokyo
0.50	11	11	11	2	1	2
0.40	17	17	17	4	3	3
0.30	39	39	39	13	7	13
0.20	55	55	55	40	37	35
0.15	117	117	117	84	85	84

Table 5.7 *Cities in cliques for 0.2 binary threshold*

City	Number of clique memberships	Proportion of total cliques
London	55	1.00
New York	55	1.00
Hong Kong	40	0.73
Paris	37	0.67
Tokyo	35	0.64
Singapore	32	0.58
Chicago	28	0.51
Milan	28	0.51
Los Angeles	19	0.35
Toronto	19	0.35
Madrid	24	0.44
Amsterdam	23	0.42
Sydney	19	0.35
Frankfurt	22	0.40
Brussels	21	0.38
São Paulo	13	0.24
San Francisco	4	0.07
Mexico City	2	0.04
Zurich	5	0.09
Taipei	6	0.11

relationship is not a smooth one. In particular, note the abrupt decline in clique membership between Brussels and São Paulo and then between São Paulo and San Francisco.
Such changes in cliquishness can be measured by expressing the total clique membership of a city as a proportion of the total number of cliques it could be a member of
(Table 5.7). However, to obtain a clearer idea of a city's cliquishness it is necessary to
go beyond such an absolute measure and measure cliquishness relative to a city's global
network connectivity.

A 'relative' measure of cliquishness can be computed by taking global network
connectivity into account in the same way as with the hinterworld analyses by using
residuals from a regression. In this case the equation is

$$y_i = a + bx_i \; (+ \, R_i) \tag{5.6}$$

where y_i is the absolute level of cliquishness for city Y and x_i is its global network
connectivity; R_i represents the residual. Residuals are departures from the cliquishness
(y_i) and its predicted level (y_i') using equation 5.6 after calibrating values for the
constants a and b:

$$R_i = y_i - y_i' \tag{5.7}$$

Positive values of R indicate that a city is a member of more cliques than would be
expected for its connectivity; negative values indicate less clique membership than
expected. Thus the residuals provide a relative measure of city cliquishness.

In applying this methodology, I have selected cities are follows. First, London and
New York are removed from the analysis; their pre-eminence is not in doubt. Here I

am concerned with other world cities. Second, I include only cities that are members of two or more cliques, thus excluding cities in no cliques but also those in just one clique; the latter constituted relatively large sets of cities, thus creating a distribution of cliquishness residuals concentrated at one low value. This selectivity resulted in different numbers of cities being analysed at different levels of the binary threshold. For instance, with the threshold at 0.3 only ten cities qualify for analysis; at 0.15 there are forty-eight that qualify. To compare across thresholds I focus upon just the five highest positive and negative residuals.

This extreme cliquishness is shown in Table 5.8. The results for the 0.30 binary threshold repeat the finding of Paris's being under-cliqued in Table 5.6 but now show it in a way that makes wider comparisons clearer: it can be seen to be by far the largest negative residual. As well as confirming Tokyo and Hong Kong's high level of over-cliquishness, the other noteworthy feature of the 0.3 threshold results is the contrast between Chicago and Los Angeles. This difference is repeated in the analysis using a 0.2 threshold. It might have been thought that Los Angeles with its Pacific connections (Figure 5.11) would be the more cliquish, but the opposite is the case here. Clearly, Los Angeles' extra-United States links are below the connectivity level being incorporated into these two clique analyses. In the second analysis (threshold 0.2) the key feature is the contrast between European and North American cities (excepting Chicago and Zurich). This is perhaps less surprising since leading US cities, in particular, have connectivity constituted to a sizeable degree by links to the numerous other US cities that appear with no cliques. Brussels' appearance as the most over-cliqued city at this relatively high binary threshold, is significant. Finally, analysis using the 0.15 threshold brings in many more cities and produces quite different extreme cliquishness. Two cities, in fact, 'change sides' in the table: Zurich and Mexico City were very under-cliqued with the 0.2 threshold, whereas they both feature as very over-cliqued in the final analysis. What is happening here is that this lowering of the binary threshold brings in large numbers of new binary links and that these two cities have a disproportionate number of their proportional city interlinks within this range. In general, it can be seen that this is a range in which cliquishness involving the leading non-core cities greatly expands. In contrast, the leading cities in the analysis are under-cliqued here: their proportional city interlinks at this level do not reflect their overall connectivity.

Table 5.8 Extreme cliquishness among world cities

Binary threshold 0.3		Binary threshold 0.2		Binary threshold 0.15	
Over-cliqued	Under-cliqued	Over-cliqued	Under-cliqued	Over-cliqued	Under-cliqued
Tokyo (0.074)	Paris (−0.097)	Brussels (0.083)	San Francisco (−0.084)	Buenos Aires (0.25)	Hong Kong (−0.20)
Hong Kong (0.043)	Los Angeles (−0.037)	Milan (0.077)	Los Angeles (−0.075)	Zurich (0.24)	Paris (−0.17)
Chicago (0.035)	Toronto (−0.028)	Frankfurt (0.072)	Toronto (−0.061)	Mexico City (0.16)	Tokyo (−0.16)
Sydney (0.028)	Singapore (−0.021)	Chicago (0.042)	Zurich (−0.058)	Mumbai (0.15)	Dublin (−0.14)
Frankfurt (0.023)	Milan (−0.019)	Taipei (0.040)	Mexico City (−0.058)	São Paulo (0.14)	Barcelona (−0.14)
				Prague (0.14)	

What do these results reveal about the world city network? The starkest finding is the fact that in terms of cliques, London and New York are everywhere. This will be explored in later analyses. The second key point is that clique results are sensitive to the binary threshold used. This means that it is not possible to make general statements about the cliquishness of cities beyond London and New York. All statements have to be prefaced by identifying the threshold level. All this adds up to a highly complex pattern of inter-city relations that cannot be easily comprehended through this standard network analysis. The remainder of the global urban analyses reported in this book attempt to reduce this complexity.

A landscape of globalization

Clique analysis divides up a connections matrix resulting in quite complex results, as just illustrated. The opposite of such analysis is a synthetic methodology that brings together results to show their relations as a relatively simple framework. Typically, such syntheses reduce complex data into a new 'space' of relations. In this section I use a standard procedure creating new spaces: multidimensional scaling (Kruskal and Wish 1978; Young 1987). The technique is 'multidimensional' in nature because it produces spaces of different orders from simple scales (one-dimensional), through two-dimensional 'maps' and three-dimensional diagrams, to 'hyperspace' models of four or more dimensions (Cliff et al. 1995: 102).

In geography, multidimensional scaling has tended to use two-dimensional solutions to create alternative 'maps' to conventional physical maps. I follow this lead here to create a two-dimensional 'global service space'. This new 'map' of world cities is then filled with a familiar content: the global network connectivities. Isolines of equal connectivity are drawn to create 'contours' across the global service space. The resulting 'contour map' describes a landscape of cities in globalization on which I apply a 'terrain analysis', a simplified description of contemporary inter-city relations.

New spaces for old: multidimensional scaling

Multidimensional scaling uses a dissimilarity matrix that measures differences between locations (Cliff et al. 1995: 104). The dissimilarities are interpreted as distances for constructing the new space. Commonly in geography, travel times between places have been used to create a time–distance matrix for this purpose. For instance, Forer (1978) used airline travel times to create a time–distance matrix for Pacific region cities to create a new 'plastic space' of the Pacific region. This two-dimensional multidimensional scaling output produces 'new maps' that are 'plastic' because they show a distortion of the physical locational map, with some places being drawn together (relatively short time-distances), while other places are pushed apart (relatively long time-distances). In the scaling reported here I use an inter-city matrix of 'global service distances' to create a 'global service space'.

To derive the global service distance matrix I return to the proportional relations matrix **Q** as illustrated in Table 5.4. The proportional city interlinks found in this matrix can be converted to global service 'distances' by subtracting them from unity:

$$d_{ab} = 1 - p_{ab} \qquad (5.8)$$

where $0 \le p_{ab} \le 1$. In this case every 'self-distance' is, by definition as a distance, given as zero. The resulting distance matrix, **D**, defines inter-city relations as low dyad values indicating high connectivity. Table 5.9 shows these distances and indicates that in 'global

Table 5.9 *Services distance matrix*

Cities:	1	2	3	4	5	6	7	8	9	10
Cities:										
1	0.00	1.00	1.00	0.44	0.44	0.93	0.44	0.93	0.89	0.81
2	1.00	0.00	0.89	0.89	1.00	0.89	0.78	1.00	0.89	0.89
3	1.00	0.89	0.00	0.97	1.00	0.97	0.93	1.00	0.97	0.97
4	0.44	0.89	0.97	0.00	0.33	0.85	0.26	0.89	0.85	0.74
5	0.44	1.00	1.00	0.33	0.00	0.89	0.33	0.89	0.89	0.78
6	0.93	0.89	0.97	0.85	0.89	0.00	0.81	0.97	0.97	0.93
7	0.44	0.78	0.93	0.26	0.33	0.81	0.00	0.89	0.81	0.70
8	0.93	1.00	1.00	0.89	0.89	0.97	0.89	0.00	1.00	0.97
9	0.89	0.89	0.97	0.85	0.89	0.97	0.81	1.00	0.00	0.97
10	0.81	0.89	0.97	0.74	0.78	0.93	0.70	0.97	0.97	0.00

Key: 1 Chicago; 2 Frankfurt; 3 Hong Kong; 4 London; 5 Los Angeles; 6 Milan; 7 New York; 8 Paris; 9 Singapore; 10 Tokyo

service distance' London and New York are the two closest cities (0.26), followed by New York and Los Angeles (0.33) and London and Los Angeles (also 0.33).

Computationally, the technique involves an iterative process in which the objects converge towards a spatial distribution of locations that minimizes a measure of 'stress' (badness-of-fit comparing distances between objects in the new configuration and in the original distance matrix). The more dimensions the less the stress, since there is more potential for reproducing the input of distances. Using just two dimensions works only when the initial data are highly structured, and fortunately this is the case with the service value matrix.

World cities: terrain and regional analyses

Although the data are highly structured, it is unreasonable to expect multiple dimensional scaling to create an interpretable two-dimensional space for an excessively large number of cities. This was found to be the case when analysing the 123 × 123 inter-city service distance matrix. There were found to be many anomalous locations of cities – cities out of place – that require the addition of extra dimensions to characterize adequately their relative positions in a global service space. However, full multidimensional studies, using more sophisticated techniques, are left to the next chapter. Here I remain committed to two dimensions, an actual mapping, so that the only solution is to reduce the number of cities. I focus on the top 62 cities in terms of global network connectivity. These all have at least one-third of London's connectivity. Additionally, they include no 'specialist' cities: Luxembourg City, the highest-ranking international financial centre with little or no additional world city functions, falls just below this cut-off point.

The new map of global service space contoured for global network connectivity is shown in Figure 5.16. Notice the location of London and New York together at approximately the centre of the map; these cities define the peak within the landscape surface. However, they appear ex-centric to the largest cluster of cities that lie to the 'east' of the peak. This is a cluster of all the leading world cities, located on a 'high plateau' below the peak. In the details of these locations it can be seen that Brussels is surprisingly 'central' compared to the more highly connected Chicago and Los Angeles. In

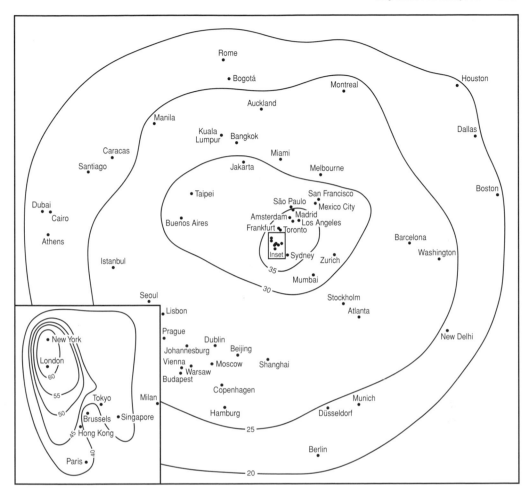

Figure 5.16 A landscape of globalization

general, this reflects the fact that it is the leading West European and Pacific Asian cities rather than US cities that constitute the dense middle of the core just 'below' London and New York.

Beyond this 'core cluster' of major world cities there is only one other comparably large and dense cluster of cities. This is to the 'south-west' and is separated from the core by a wide city-less 'plain'. The cluster features those world cities beyond the top echelons that act as gateways to so-called 'emerging markets'. Thus it includes Chinese and East European cities. This reflects global strategies of firms that have taken particular advantage of the 1990s phase of new expansions of globalization (which is why Dublin and Johannesburg also appear in this cluster).

Beyond these two large clusters the remaining cities lie in regional patterns about the core cluster. To the 'west' the terrain falls abruptly with no cities near to the peak. This hemisphere (which includes the 'emerging markets' cluster) includes mostly what were 'Third World' cities. Here they are distributed into three 'tongues' pointing out from the centre: Pacific Asia (from Taipei and Jakarta outwards), Latin America (from Buenos Aires outwards) and Middle East/eastern Mediterranean (from Istanbul outwards). On the slopes to the 'east' of the core cluster it is Western cities that predominate. There is some further patterning here such as relatively minor US cities in a 'north-west' arc and German cities taking up a 'southern' position. Overall, however, using just two dimensions the city regionalism is largely reflected in a broad 'Third World/First World' division with the 'emerging markets' cluster in the former and the core cluster in the latter.

What has this synthetic analysis achieved? There are two features that stand out. First, the new map provides hints about how cities fit together in the world city network. There appear to be both hierarchical and regional tendencies in the new terrain. This is explored much further in Part III, where the focus is upon configurations of services and cities. Second, by producing a new 'worldwide map' of cities, this synthesis has provided a fresh way of looking at world cities and their inter-relations. This is an elementary contribution to visualization of world cities in globalization. This theme is taken up in a brief concluding discussion.

Visualization of the world city network

In Chapter 2 I used Thrift's (1999: 271) description of the contemporary space of flows as a 'blizzard of transactions' to indicate the complexity of this book's subject matter. The rest of the book, both before and after the current discussion, can be interpreted as attempts to steer a way through the blizzard to reach some sort of understanding of the complexity. The previous section's unusual 'terrain analysis' is just the latest of several unconventional ways in which results are presented to help in illustrating findings. The cartograms represent another effort to show analytical results in a simple and direct manner. These are all elementary examples of visualization techniques.

Visualization is about creating easily interpretable pictures from complex data (McCormick *et al.* 1987; Orford *et al.* 1998). In the case of world cities this means converting their myriad relations into intelligible images of a new world-space. Displaying world cities on uncomplicated maps represents only a first stage in visualization for a new metageography, however. Visualization is about much more than image creation; data are not just passively displayed, they are depicted in such a way as to generate new knowledge. In visualization the image is not the final output; the viewer interacts with it to find new patterns and relations in the data. It is this feature that makes visualization potentially so important: to go beyond display and invite people to think in new ways about their world so as to combat the dominance of a mosaic perception of world-space, a prime task of this book.

However, the difference in visual complexity involved in displaying spaces of flows compared with maps of spaces of places is quite staggering. Simple choropleth maps of differentially shaded areas – so common in world development studies – provide clear and simple depictions of their subject matter. In contrast, flows are notoriously difficult to portray in simple cartographic form. There is a limit to the number of flow-arrows that can be portrayed before overlapping creates a confused image. Such an image can work as a visualization device when the purpose is to express the magnitude and complexity of flows – Thrift's (1999) 'blizzard'. But the detail is lost in the mass of information portrayed. Evidence from the recent *Review of Visualization in the Social Sciences Report* (Orford *et al.* 1998) suggests that in the social sciences at least, there have been few attempts to create visual models of complex flow patterns. There are examples of economists attempting to portray trade patterns (p. 26) and sociologists doing the same for social networks (p. 27), but the conclusion of the report is that social scientists are only at the beginning of this work.

The key problem with the visualizations above, and the examples Orford *et al.* (1998) describe, is that they are all essentially static in nature. Since flows are intrinsically mobile in nature, it is at least plausible that quite different media will ultimately become the way in which spaces of flows, including that world-space of flows organized through world cities, will be visualized. Orford *et al.* (1998) point towards animation and virtual reality, with the World Wide Web being the 'ideal medium' (p. 12), as the way forward. Animations provide for the possibility of showing flows between cities as movement. Clicking on a city and seeing its inter-city connections moving across the screen is the starting point here. Virtual reality techniques hold the prospect of constructing corporate service space and allowing an interactive traveller to be a 'message', moving around the space to explore its structure while 'contributing' to new city knowledges. The pedagogic implications of this are quite exciting. But they cannot be followed up in the medium of this current report of researches, being as it is in 'old-fashioned' physical book mode.

Part III Configurations

6 ● A mapping of services in globalization

> Globalization takes place in cities and cities embody and reflect globalization. Global processes lead to changes in the city and cities rework and situate globalization. Contemporary urban dynamics are the spatial expression of globalization, while urban changes reshape and reform the processes of globalization.
>
> (Short and Kim 1999: 9)

The final results of Part II used multidimensional scaling to depict cities in a two-dimensional space. In Part III more of the complexity within the data is described by allowing the analyses to include more than two dimensions. To achieve this end a related but more sophisticated technique is used: principal components analysis. As a member of the factor analytic family of techniques, this new analysis maintains a synthetic purpose in that it searches out common patterns within the data. Thus inter-relations both among the service firms and between the cities are analysed in order to delineate spatial configurations in a series of global urban analyses.

There are two types of analysis I conduct on the services and cities data. The matrix can be analysed either by columns to focus upon relations between firms or between rows to focus upon relations between cities. The former type of analysis reveals configurations among the firms, showing similarities in distributions of offices, and is the subject matter of this chapter. The latter reveals configurations among the cities in terms of similarities in their service mixes and is the subject matter of Chapter 7. By searching out patterns in this way, spatial configurations are identified that show different geographies of globalization.

The analyses in this chapter produce a first comprehensive global mapping of advanced producer services. The argument is divided into three sections. I begin by rehearsing key considerations for studying services under conditions of contemporary globalization. From this I suggest what geographies might be expected for globalized services. The second and third sections are the core of the chapter – reporting the analysis – through which I assess these expectations empirically. In fact, I produce a set of six quite different patterns of global locational strategy among service firms. These multiple globalizations are generally related to a firm's service sector, different world regional opportunities and constraints, and the rankings of cities. In the final section I return to the media data introduced in Chapter 5, to compare global service configurations with global media configurations.

Between them, the latter comparison and the preceding main analysis should put paid to any idea that globalization is a singular process that is transforming the world economy into a simple, single economic geography. Globalization is so complex, even within the narrow empirical confines of the analyses conducted here, that the thought kept crossing my mind on viewing the results that there should be a campaign to make 'globalization' one of those nouns that is always referred to in the plural. Certainly, the watchwords of this chapter are 'multiple globalizations'.

Towards new geographies of services

Traditionally, study of the geography of services has been where urban geography and economic geography have intersected through central place theory. This abstract hinterland model treated cities as 'central places' servicing their surrounding local region. The latter varied in size, incorporating different geographical scales of hinterland to create urban hierarchies. The 'national urban systems' I described in Chapter 1 are the culmination of this type of modelling. As previously noted, it resulted in a systemic bias in the treatment of great cities such as New York, Paris and London. They were invariably viewed as 'national', the largest scale of hinterland, rather than 'international', a dimension of activities in these cities typically neglected. This nationalist manifestation of prioritizing an 'internal' over an 'external' pattern of relations was changed only in empirical studies of cities as trans-shipment points for commodities (e.g. seaports) where the national 'internal' hinterlands are empirically balanced by descriptions of international (i.e. external) 'forelands'.

The hinterworlds I described in the previous chapter may be considered a latter-day combination of hinterlands and forelands for new information commodities. The key idea I have added is the treatment of cities as part of a carefully specified network. And, of course, forelands were not usually about services. Today, services are considered to be a commodity but they have traditionally been perceived as something quite static. Surely such an intangible economic transaction as providing and receiving a service in a given place cannot be a traded commodity.

The myth of non-tradeability of services

The business services I am concerned with here are financial and producer services that provide high-skill, knowledge-based inputs into the production process of other firms. These services are typified by intensive provider–client interaction so that specialist knowledge can be converted into customized products (Aharoni 1993a: 8–9). Traditionally, such services have relied on face-to-face contacts to build up client lists that are quite local. For instance, law firms have long been identified in terms of the city in which they practise: a Boston law firm, a New York law firm, a Birmingham law firm, a London law firm, and so on. This situation led to the idea that such services were essentially 'location-bound' (Roberts 1999: 73). However, some of these services have long broken out of local markets to become national in scope, notably in banking, and subsequent 'internationalization of services' has finally ended the myth of the non-tradeability of services.

Aharoni's (1993b) influential edited collection of studies on 'the globalization of professional business services', and the recent reviews by Bagchi-Sen and Sen (1997) and Coffey (2002) of the new geographies of producer services, between them provide an indication of the key themes that have concerned researchers in this field. Broadly speaking, this work concentrates on the supply and demand factors that have led to service firms providing their wares in more than one country. The many impediments from national regulations of professions (critical in law) to cultural differences (critical in advertising) are highlighted to indicate the power of the 'globalization drivers' that have overcome such obstacles. Key drivers commonly identified include many of the topics described in earlier chapters: first, the changing nature of the economy, whereby service firms have to follow clients that produce in many countries or else lose their custom; second, the liberalization of the world economy, creating myriad servicing opportunities; and third, enabling information and communication technology that makes such non-local service provision possible. Typical specific studies of this process include investigation of the determinants of international investments in services (e.g. Li and Guisinger 1992), the corporate strategies and modes of delivery (e.g. Vandermerwe and Chadwick 1989), and identification of stages in the process (Edvardsson *et al.* 1993; Roberts 1999).

Roberts (1999) identifies five stages in the 'internationalization of business service firms'. Starting with purely domestic clients and therefore no service exports (stage 1), domestic practice extends to servicing foreign clients within the country (stage 2) before service exports begin in 'embodied' form with practitioners visiting countries where clients have business (stage 3). The last two stages constitute the globalization that is my concern here. In stage 4, firms expand their office network to have a presence in foreign markets to which they export through intra-firm flows from domestic to foreign offices. Finally, in stage 5, services are produced in foreign offices as well as domestic ones. Stage models can be dangerous tools because they imply an inevitability in a process, but as long as they are employed for pedagogic purposes they can be very useful in clarifying practices. This is how I employ Roberts' model for dealing with the locational strategies of firms that are operating in the fourth and fifth stages. For instance, I have shown that law firms define the least globalized service in the sense of having offices in fewer cities across the world. However, because the data I use are limited to just global law firms (in fifteen cities by definition), they miss out the 'foreign strategies' of many other law firms. In one study incorporating the largest 250 US law firms (Beaverstock *et al.* 2000b), it was found that only 100 actually had foreign offices. And of these, over 50 per cent were located in just one or two foreign cities (p. 106). Furthermore, it was noted that US global law firms are not necessarily the most profitable US law firms (ibid.: 116). As Spar (1997: 14) comments, 'the emergence of global law has not made all firms global. Nor has it made all global firms successful.' Thus there is no clear-cut reason for all other major US law firms to 'globalize', an expensive business in a professional service. This means that there is no necessary pattern of law firms moving through stages in the Roberts manner, although those that are in the service and city data used here might well have done so. Finally, in general I should note that the final two stages that are of interest here imply different globalizations: direct foreign service production suggests more commitment to a globalizing strategy and may mean having many more offices across the world than is necessary for simple intra-firm exports. But it will not be that simple; stages will interact with service sectors to produce a variety of results. Roberts (1999: 69) does admit that his subject matter 'is yet to be fully explored'; some new geographical explorations are detailed below.

New economic geographies

Although globalization is inherently geographical in nature, the geography of the locational spread of business service provision is only weakly represented in the subject overviews referred to above. Bagchi-Sen and Sen (1997) do provide tables showing the home countries of leading accountancy and advertising firms and how many countries they operate in, but there is no geography discussed beyond very broad categories such as 'Asia-Pacific' and 'Europe' (table 5, p. 1171). However, there is a literature on the new geography of services, notably in the pioneering work of Daniels (1993) and Daniels and Moulaert (1991), which is reviewed in Dicken (1998: chapter 12). Such work has included inter-regional comparisons within the 'supply countries' (e.g. O'Farrell *et al.* 1996) and numerous studies of particular sectors including banking/finance (e.g. Corbridge *et al.* 1994), advertising (e.g. Leslie 1995), law (e.g. Beaverstock *et al.* 2000b) and accountancy (e.g. Beaverstock 1991).

These new economic geographies provide detailed glimpses of the globalization of particular service firms but it is very hard to draw any general conclusions from them. Clearly, a few important cities are regularly identified as important, but there is no overall world city network context in which to interpret these findings. The large services and cities data set provides the opportunity to go beyond particular patterns, and its antecedent network specification provides a conceptual context in which to view more comprehensive empirical results.

Multiple globalizations

The geographies of firms briefly described above are, of course, all different. This is what would be expected; describing patterns at the level of the individual decision-making unit inevitably creates a variegated geography. Equally, if such patterns are combined and focus is transferred to their common features, as happens in any statistical analysis, there will be a tendency to suggest a homogeneity, perhaps an average pattern that is called upon to play the role of a representative ideal. Certainly, much globalization literature suggests that the latter might actually fit the reality. But this is to see globalization as an end state rather than as a bundle of processes (Taylor 2000b). This is because the concept of globalization has been built upon many myths, the most notable being that the world has become homogenized under the relentless pressure of global capital. This abstract capitalism is deemed to have descended upon the contemporary world like a huge blanket, eliminating geographies through instantaneous communications, creating a uniform conformity where once there had been diversity. There is an urgent research need to escape from the blanket but without recourse to the alternative comforter of insisting on singularity.

Both positions need critiquing. First, the changing scale of much economic activity (to become effectively global) does not necessarily result in a homogeneous world. As Dicken (1998) has taught us, economic globalization has not simply 'emerged'; it is the creation of agents, of large corporations aided and abetted by neo-liberal state policies. In the case of financial and business service firms, locational strategies are marked by geographies of different national backgrounds and histories of different sectoral expansions. In aggregate this is a bundle of many different global processes creating many different geographies. The result has been the creation of very complex patterns of globalization consequent upon globalizing agents coming from a variety of different types of economic contexts. But I must be careful not to go too far with this argument. Second, although every firm's pattern of globalization will be unique to that firm, these patterns are not isolated and singular: they are each subject to the same general drivers and specific attractions. Some of the latter are very obvious: all banks with any global pretensions will have a London office, all advertising agencies with a global vision will have to be in New York, and all service firms with their sights on the potential of China's market will have to seriously consider locating in Hong Kong. For all firms there will be many such pressures, some sectorally specific, others geographical in nature, and still others relating to continuing national patterns, as illustrated in the national breakdowns of global network connectivities in Chapter 5. Thus I can expect similarities among firms in the same sector or from the same country/region or responding to the same drivers.

What does all this suggest? Neither simple singularity nor simple homogeneity, but rather a complexity within contemporary globalization. Common attractions and constraints in firms' decision making will combine with their myriad distinctive expansions to create what I term 'multiple globalizations'.

Defining common global patterns of service provision

Common patterns of services are derived from the data described in Chapter 3. As with previous analyses, the focus is on the top 123 connected cities. Hence the data matrix I work with is a 100 firms × 123 cities matrix of service values ranging from 0 to 5. Results will be displayed on the same cartogram as is used to illustrate connectivities (Figure 4.1).

The data matrix is interpreted conventionally as firms being variables and the cities being objects. In this way each column of the matrix is a 'firm as variable' consisting of 123 service values. The distribution of these values across the 'cities as objects' represents the global locational strategy of the firm. Each column will be different because no two firms have exactly the same global locational strategy. Thus the matrix describes the distinctiveness of firms. But it also shows similarities between firms. Large numbers of relatively high correlations between the columns show that the matrix is most certainly not the random collection of service values that would indicate singularity of firms. Such similarities indicate that there are common patterns among the firms, and I build upon this to define those patterns.

Parsimony through principal components

The task is clear: faced with a matrix that contains 12,300 (100 × 123) pieces of information, I need to reduce the detail into a relatively small number of common patterns for interpretation. Such parsimony is the basic purpose of the factor analytic family of statistical techniques, of which the most straightforward (i.e. with least axiomatic baggage attached) is principal components analysis. This is the technique I apply to the matrix of service values.

In principal components analysis, a data matrix consisting of x variables is treated as an x-dimensional space to which each variable contributes an axis. Each axis is one unit in length, which represents the spread of values (variance) of a variable, so the total variance equals x, the number of variables in the matrix. Analysing the co-variance (correlation) among the variables makes it possible to reorganize the axes so that different-length axes are produced within the x-dimensional space. These new axes are created in order of size (their length) and are termed the principal components of the original variable-space. In this analysis the key property of this rearrangement of the space is that the first axis (largest component) describes the biggest cluster of the original variables as reflected in their co-variances (correlations), the second component the next biggest cluster of variables, and so on down to a very small final xth component that accounts for hardly any of the variance in the data. The basic idea is that most of the variation in the data is revealed in a relatively small number of large components (y), so that many small components can be discarded as unimportant. The data transformation is interpreted, therefore, as a conversion of an original x-dimensional variable-space into a much smaller y-dimensional component-space defined by selection of only the large and important principal components. This is the parsimony, converting a large number (x) of variables into a relatively small number (y) of important components.

The fundamental test of the veracity of the technique is found in the proportion of the original variance, x, that is described by the y components. Obviously, if all components are selected ($y = x$), then the variable-space has been reordered but not reduced, so that 100 per cent of the original variance is described. However, since the original variance is concentrated in the higher-ranked components, it will be found that, say,

29 per cent of the original variance is described by the two largest components. Although impressive, this would normally be viewed as a partial analysis because much less than half of the variation in the data is accounted for. Typically for large data matrices, results of a principal components analysis will consist of between four and nine components accounting for between 50 and 70 per cent of the original total variance. In the analysis reported below, the 100 firm-variables are reduced to just six components that account for 52 per cent of the variance in the matrix of service values. With 100 different firm patterns reduced to just six common patterns, this is indeed a very parsimonious description of a very large data matrix.

Extracting principal components from a data matrix is a standard statistical procedure; the skill lies in the interpretation of the components. The chief way of interpreting components is to focus upon the location of these new component axes with respect to the original variable axes. In the x-dimensional space the new axes are located within clusters of the original variable axes, and the closeness between old and new axes can be measured. These measures are called 'loadings' and are interpreted as correlations between a variable and a component. Thus a high loading, say 0.9, means that the variable and component are very close together in the variable-space. Because variance is concentrated in just a few components, it follows that each component will be expected to have a number of variables that load high on it. Hence components represent clusters of like variables in the variable-space: by looking at the high loadings/correlations, the cluster of variables that a component represents is revealed. In order to facilitate this interpretation, the principal components are usually 'rotated' so as to maximize high loadings. The most common method is called 'varimax rotation', which creates well-defined and orthogonal (independent) patterns of variability as components. It is these rotated principal components that are used below for a parsimonious analysis of the data matrix of service values. In effect, these rotated components are new 'composite variables' and can be identified and named from inspection of the high-loading variables.

There is one crucial decision that has to be made in any principal components analysis: the number of relevant components (y) to be selected from the x number of components that are actually created. There is no simple statistical answer to this question; ultimately it comes down to the researcher's judgement. The question is explored in some detail in Chapter 6. In the analysis below, the six-components solution is selected because it provides a particularly interesting set of common patterns. In addition, with six components as the outcome, the possibility that each of the six service sectors in the data could form their own component-cluster of firms is left open. In fact, as will be seen, sectors turn out to be very important in the analysis, so just such an eventuality almost happens.

Component-clusters and service sectors

The results of the principal components analysis are shown in Box 6.1. All loadings above 0.4 are included and firms are ranked by the size of their loading under each component. It is immediately clear from this table that the clusters of firms are closely related to the different service sectors. This has made naming the components an easy task; in Box 6.1 firms are highlighted in the sector(s) used to label each component. However, the component–sector relation is not an exact one: management consultancy, advertising and accountancy each have their own components but the make-up of the other three is a little more complex. Insurance, the sector with least firms included in the data, is not represented by its own component. Instead, two sectors share a component (law and banking/finance), while also having their own distinctive components, thus making up the total of six components. I describe the two groups of components in turn.

Box 6.1 Component loadings for global service firms

Component I: Law and banking/finance

0.81 **Bayerische LG** (BF); 0.78 **Freshfields BB** (LW); 0.74 **Sakura** (BF); 0.73 **Lovells** (LW); 0.72 **Linklaters** (LW); 0.71 **Clifford Chance** (LW); 0.67 **Cameron McKenna** (LW); 0.66 **Commerzbank** (BF); 0.62 **Allen & Overy** (LW); 0.60 **WestLB** (BF); 0.58 **Deutsche** (BF); 0.58 Allianz (IN); 0.55 **Dresdner** (BF); 0.55 Winterthur (IN); 0.52 Nexia (AC); 0.52 **White & Case** (LW); 0.49 AGN (AC); 0.48 PricewaterhouseCoopers (AC); 0.48 McKinsey (MC); 0.47 Cap Gemini (MC); 0.45 Sema (MC); 0.44 **Coudert Brothers** (LW); 0.44 Boston (MC); 0.43 Fortis (IN); 0.42 **ABN** (BF); 0.40 **Fuji** (BF)

Component II: Management consultancy

0.69 **Andersen** (MC); 0.67 **Hewitt** (MC); 0.67 **Kearney** (MC); 0.64 **McKinsey** (MC); 0.63 **Watson Wyatt** (MC); 0.61 **Towers Perrin** (MC); 0.59 Chubb (IN); 0.56 **Bain** (MC); 0.55 CMG (AD); 0.55 TPM (AD); 0.54 **Boston** (MC); 0.52 Bank of Tokyo (BF); 0.51 Arthur Andersen (AC); 0.50 **Booz Allen & Hamilton** (MC); 0.48 Reliance (IN); 0.42 Impiric (AD); 0.40 Chase (BF); 0.40 Jones Day (LW)

Component III: Banking/finance

0.79 **J P Morgan** (BF); 0.76 **HSBC** (BF); 0.76 **Sumitomo** (BF); 0.76 Asatsu (AD); 0.75 **Bayerische HV** (BF); 0.70 Hakuhodo (AD); 0.61 Prudential (IN); 0.59 **Dai Ichi Kangyo** (BF); 0.55 **Sanwa** (BF); 0.49 **Fuji** (BF); 0.47 **Chase** (BF); 0.47 **UBS** (BF); 0.47 **Citigroup** (BF); 0.44 **Barclays** (BF); 0.42 **Bank of Tokyo** (BF); 0.41 CGNU (IN); 4.0 Baker and McKenzie (LW)

Component IV: Advertising

0.74 **Ogilvy** (AD); 0.72 IBM (MC); 0.71 **D'Arcy** (AD); 0.70 BDO (AC); 0.68 **McCann-Erickson** (AD); 0.67 **FCB** (AD); 0.66 Deloitte Touche (MC); 0.63 **Young & Rubicam** (AD); 0.60 Grant Thornton (AC); 0.60 **Saatchi and Saatchi** (AD); 0.56 **J Walter Thompson** (AD); 0.53 **BBDO** (AD); 0.52 **Impiric** (AD); 0.43 BNP (BF); 0.40 Nexia (AC)

Component V: Law

0.72 **Latham & Watkins** (LW); 0.70 **Morgan Lewis** (LW); 0.69 **Skadden, Arps, etc**. (LW); 0.67 **Morrison & Foerster** (LW); 0.67 **Sidley & Austin** (LW); 0.62 **Dorsey & Whitney** (LW); 0.55 **Jones Day** (LW); 0.53 **Coudert Brothers** (LW); 0.49 **White & Case** (LW); 0.48 Mercer (MC); 0.47 CSC (MC); 0.42 Chubb (IN); 0.41 Towers Perrin (MC); 0.40 PricewaterhouseCoopers (AC)

Component VI: Accountancy

0.62 **PKF** (AC); 0.57 Lloyd's (IN); 0.53 **MacIntyre Strater** (AC); 0.53 CSC (MC); 0.47 **IGAF** (AC); 0.46 **Howarth** (AC); 0.45 **Summit** (AC); 0.45 **HLB** (AC); 0.45 Royal & SunAlliance (IN); 0.44 **AGN** (AC); 0.41 CGNU (IN)

(Firms from leading sector in each component are emboldened)

Management consultancy, advertising and accountancy

Components II, IV and VI are three straightforward components as far as labelling is concerned. Component II is the most clear-cut sector-component of the three, with nine management consultancy firms featured, including the highest six loadings. This contrasts with accountancy (component VI), with just seven of the eleven firms featured. The last figure reminds us that this is the least important of the components despite its being named after the leading sector in terms of overall connectivity. Accountancy firms not listed on component VI are scattered across all other components except III. This reflects accountancy's ubiquitous nature, as discussed previously. Advertising is between these two examples, featuring nine firms in component IV but with not all so highly placed as management consultancy firms are in component II.

These three components show that there is a strong tendency for firms in the same sector to have similar distributions of service values across world cities. This would seem to be the result of distinctive sector origins/histories and different sector opportunities in global sector markets that have arisen within contemporary globalization. Further evidence of this is given in the next section. However, the message of the analysis so far could hardly be clearer: sectors matter in the globalization of advanced producer services.

Banking/finance and law

The importance of sectors does not diminish when components I, III and V are considered, but the picture is slightly more complicated. There are two clear-cut service-components. Component V is unmistakably a law component, its identity being even more explicit than management consultancy in component II: nine law firms are featured and they constitute the top loadings on the component. These represent nine of the ten US law firms in the data; the exception is Baker & McKenzie, the largest US law firm, which has previously been shown to be distinctive among US law firms (Beaverstock *et al.* 2000b). Component III is also quite straightforward since it features just over half the banks in the data (12), including the three firms loading highest on the component.

This leaves component I as the odd one out, the component that prominently features more than one service sector. There are eight each of banking/finance and law firms featured in Box 6.1 on this component and they share the highest eleven loadings, six for law, five for banking/finance. It is at this top level that the nature of the component becomes clear. The six law firms at the top of the list constitute all the London law firms in the data. In addition, German banks constitute six of the seven highest-loading banking/finance firms. In law this reflects the recent expansion of London law firms into Germany, involving several mergers. In banking it is German banks using London as their base for the global business arm of their operations. Thus at the heart of this component is the expansion of London lawyers into Germany and German bankers into London. These may be two different processes producing very similar patterns of service values or there may be an evolving synergy heralding the development of a new bi-sector process. Further research is required in order to choose between these alternative interpretations.

And what of the smallest sector in the data, insurance with its eleven firms, which does not have its own component? In fact, the results for insurance firms point to factors other than sectors determining global patterns of services; in particular, the geographical origins of firms are emphasized. Insurance firms split between components as follows: non-UK European firms feature in component I, US firms are in components II and V, and London firms are to be found in components III and VI. Of course, consideration of firms' origins was also discussed in the unravelling of the law and banking/finance components. However, the insertion of this geographical element into the analysis has been very weakly developed thus far. This is easily remedied.

Alternative spatial-strategic emphases in service globalizations

The six components represent clusters of firms with similar distributions of service values across cities. Thus far, I have focused upon component loadings because they relate components to firms, as just illustrated. However, loadings are not the only results that are derived from a principal components analysis. Just as each firm (variable) displays a pattern of service values across cities, so also do the clusters of firms that are components (composite variable). In general, this can be thought of as patterns of 'composite service values' across cities, and these are computed as component scores. The latter are derived for each city across all components from the service values of firms and their loadings on each component.

There is a component score for every one of the 123 cities for each of the six components. This means that I can map each component to reveal its geography. As with loadings, I concentrate on the higher results, and in this case I use a simple mapping protocol to show the sector-components as 'service fields'. Component scores can be both positive and negative, and the positive scores on a component are used to delimit the broad dimension of a component's service field. Cities with positive scores are those that have more service firm presences in the data for firms that contribute positively to the component (i.e. their firms' loadings on the component generally show positive correlations). From among these I identify cities with scores above 1.0 as primary field cities, with the rare examples of scores over 3 indicating an 'articulating city'. In the latter case, all, or nearly all, the key firms that cluster to produce the component have important offices (service values of 3, 4 or 5) within this city (or cities). Such cities represent critical network hubs within the global service markets that the service fields depict. Nine such articulator cities are identified.

The results I present here describe six different geographies of globalization. It is important to remember that all the firms, by definition, have global strategies, so that the differences between them are matters of emphasis, such as a particular regional concentration. Thus the common patterns I describe in what follows show relative importance, where groups of firms have unusually important offices relative to the overall patterns of variation. The mappings should not be interpreted, therefore, as delineating absolute patterns of service; rather, they show common 'spatial-strategic emphases' in the global strategies of a set of firms. Because the principal components reflect the service sectors, the service fields illustrate different spatial-strategic emphases in the globalization of different sectors. Discussion of these fields is structured around the geographical locations of their articulator cities: three fields are organized through US cities, two through West European cities, and one through Pacific Asian cities. I take them in reverse order.

The Pacific Asian city articulated service field

Tokyo/Bangkok's Pacific Asian spatial-strategic emphasis (component III)

Figure 6.1 shows that this banking/finance component has a very strong regional emphasis in its service field: Tokyo and Bangkok are the articulator cities and every one of the other Pacific Asian cities features as a primary field city. It is interesting that

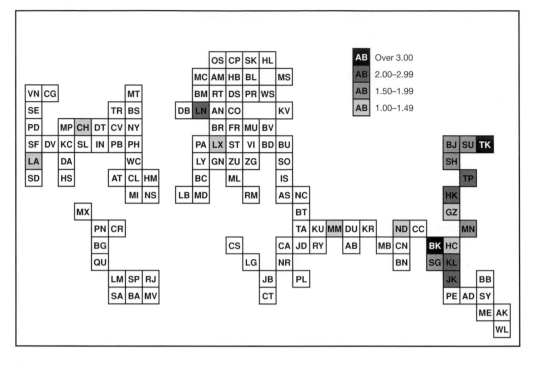

Figure 6.1 Tokyo/Bangkok's Pacific Asian service field (from Taylor *et al.* forthcoming). (For city codes, see Figure 4.1.)

Hong Kong and Singapore, although important primary field cities, are not articulators instead of Bangkok. In fact, it will be shown that these two important Pacific Asian cities feature in the other banking/finance component: the banking/finance prowess of Hong Kong and Singapore is spread between two banking service fields, implying a broader orientation in the geography of banks operating in these two cities.

This most regional of service fields is definitely not a 'Pacific Rim' phenomenon: no Australasian cities are primary field cities, therefore there is no 'eastern' Pacific Rim, and the only non-Asian Pacific Rim representative is Los Angeles. The other primary field cities are mainly financial centres such as Manama, Luxembourg and Chicago. Interestingly, London is an important primary field city in this field but the two other leading non-Asian financial centres, New York and Frankfurt, are not.

This service field relates directly to several other findings in the empirical analyses. For instance, in Chapter 4 Pacific Asian cities are shown to be particularly important for banking/finance network connectivities (Table 4.6). Here these inter-city relations are reordered as part of an over-spatial configuration of the world city network.

European city articulated service fields

The two European city articulated fields are very different from one another, with one having a definite regional core and the other expressing a more diffuse 'Westernness'.

Frankfurt/Munich's pan-European spatial-strategic emphasis (component I)

This banking/finance and law component has a very distinctive service field linking key primary field cities in the former Soviet realm (Moscow, Prague, Warsaw and Budapest) with primary field cities in Western Europe (Figure 6.2). These include all German cities in the data, with Frankfurt and Munich taking the role of pan-European articulators. Outside Germany, London and Paris are the two leading West European primary field cities. This is a pan-European pattern in the traditional sense of linking east and west; both North (Scandinavian) and South (Mediterranean) European cities (with the exception of Milan) are not to be found among the primary field cities.

Outside Europe there are only three primary field cities but they do add a global dimension to the field. Each non-European primary field city provides a link to an important world region: New York to the United States, Hong Kong to China, and Singapore to South-East Asia.

What this service field strongly suggests is that London law firms' expansion into Germany and German banks' growth in London do *not* represent a European Union economic integration (this would involve a more north–south orientation). Rather, this field reflects post-Soviet sector financial and legal opportunities that have created a service field showing a much broader European economic integration.

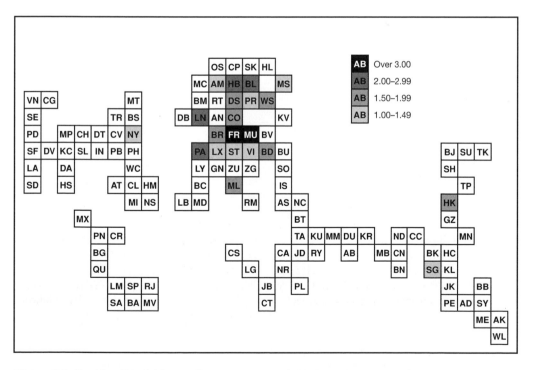

Figure 6.2 Frankfurt/Munich's pan-European service field (from Taylor *et al.* forthcoming). (For city codes, see Figure 4.1.)

London's Anglo-American spatial-strategic emphasis (component VI)

London records the third highest component score in the analysis (4.3), showing that this accountancy component's field is very strongly articulated through this single city. The service field it articulates largely consists of an Anglo-American realm (Figure 6.3). Beyond London, the primary field cities are dominated by 'Old Commonwealth' connections, with three Canadian and four Australian cities featuring prominently as primary field cities of the field. New York is another prominent primary field city, and there are four other US cities that qualify as primary.

Outside this Anglo-American realm, other primary field cities are all from Europe, with Paris prominent and including three Scandinavian primary field cities.

Although I have often referred to the ubiquitous quality of accountancy, this analysis shows that once accountancy is viewed from within the overall global configuration of services, it is in fact an advanced producer service with a distinctive Western bias in its spatial-strategic emphasis.

US city-articulated service fields

The service fields articulated through US cities are again very different, with only one of them being a primarily US service field. I describe this one first.

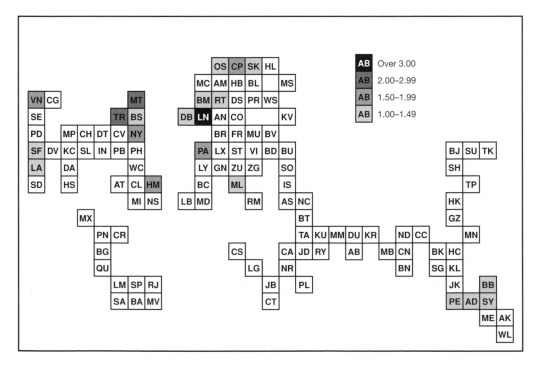

Figure 6.3 London's Anglo-American service field (from Taylor *et al.* forthcoming). (For city codes, see Figure 4.1.)

Chicago's United States spatial-strategic emphasis (component II)

With Chicago as the articulator, eight other US cities feature as primary field cities, thus making up half of the cities with component scores above unity (Figure 6.4). With management consultancy being the archetypal American producer service, this geographical concentration is not surprising. However, there is a second, less predictable feature of this service field: its primary field cities are largely the less important cities within and without the United States. Thus New York and Los Angeles in the United States and London, Paris and Tokyo beyond are conspicuous by their absence as primary field cities. And, in addition, no Pacific Asian cities feature as primary field cities.

In summary, this is a United States-centred strategy with another emphasis on important but not leading world cities outside Asia. This is the collective emphasis of US management consultancy firms in their globalization.

New York/Washington, DC's global city spatial-strategic emphasis (component V)

This legal service field has two unusual features (Figure 6.5). First, the two articulator cities, New York and Washington, DC, have by far the highest two component scores in the analysis (5.6 and 5.2 respectively). This powerful articulation confirms these two cities as the US global law cities. Second, there are relatively few primary field cities: just eight equally divided between US and non-US cities. The US primary field cities are all western, perhaps compensating for the eastern US articulation.

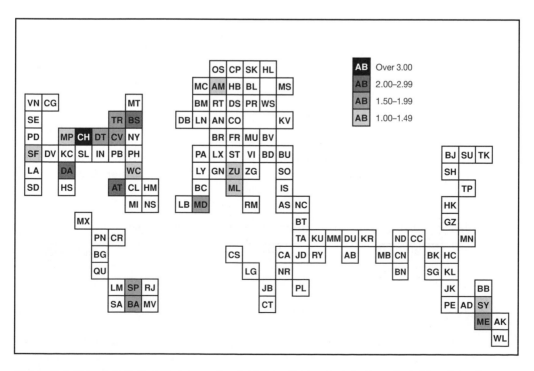

Figure 6.4 Chicago's United States service field (from Taylor *et al.* forthcoming). (For city codes, see Figure 4.1.)

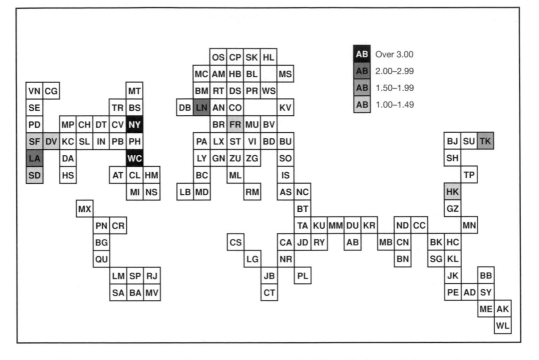

Figure 6.5 New York/Washington's global city service field (from Taylor *et al.* forthcoming). (For city codes, see Figure 4.1.)

However, the non-US cities are all very important world cities – the converse, in fact, of Chicago's articulated field just described (Figure 6.4). They consist of the two leading international financial centres in each of the two non-US globalization arenas: London and Frankfurt in Western Europe, and Tokyo and Hong Kong in Pacific Asia. Given the inclusion of Los Angeles and San Francisco among the US primary field cities, this service field can reasonably be described as having a global city emphasis.

In general, the small number of primary field cities does support previous analyses that show law firms to be the most fastidious users of cities among the six service sectors. More particularly, unlike the other law component (component I), this one is not a joint cluster with banks, but its service field does show that US law firms are using financial centres as the cornerstone for their globalization.

New York's European 'minor primate' spatial-strategic emphasis (component IV)

This advertising service field is the most unusual in the analysis (Figure 6.6). Most of the primary field cities scoring high are European (11 out of 18), and yet the field is articulated not by a European city but by New York. Furthermore, there are no primary field cities from Europe's four largest economies: Germany, France, the United Kingdom and Italy. What are recorded are minor '*primate* cities' across the continent from Istanbul to Lisbon, and from Helsinki to Athens. Elsewhere there are similar leading cities in medium-sized economies such as Johannesburg, Manila and Buenos Aires. Miami is the only US primary field city in the field, but appears here in its role as 'capital of the Caribbean/Central America' (Brown *et al.* 2002). This is a service field constituted by

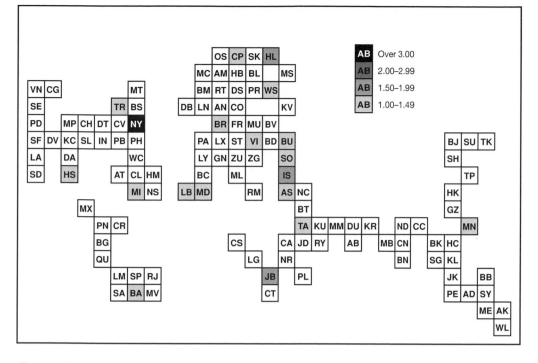

Figure 6.6 New York's European 'minor primate' service field (from Taylor *et al*. forthcoming). (For city codes, see Figure 4.1.)

medium-sized national markets. The reason advertising has to be organized this way is that global campaign implementation has to take into account the cultural and language specifics of national markets. Advertising firms are therefore over-represented in many relatively minor cities that house national television stations. Like the other services, advertising is located in major world cities but its particular geographical bias is to be found in its need to cater for national markets.

There is double irony in the peculiarities of this service field in relation to globalization. First, advertising is a service with a global image and global structure and yet it does not transcend national markets in its service field. Second, its service field exhibits the most transnational feature: it is articulated extramurally through a global service leader (Madison Avenue).

Sectors, regions and ranks

What this analysis has shown is that globalization of service provision is very complex. There are multiple globalizations created by the interaction of three distinct influences: the characteristics of different service sectors, the attractions of different regions, and the rankings of cities in terms of importance. The significance of sectors to the way service firms have devised global locational strategies is clearly illustrated in the previous subsections. In this subsection I have added region and rank as attributes of cities that are also important.

Even though the analysis includes only global service firms, the resulting geographies are remarkably regional in nature. Global service provision does not create an even world regional distribution. It is the main 'globalization arenas' (northern America,

Western Europe and Pacific Asia) that dominate: they house all nine articulator cities and the vast majority of primary field cities in all service fields. Pacific Asia is the most distinctive region in this analysis, with its own service field (component III). Northern America and Europe each dominate two service fields (components II and V, and I and IV, respectively) and they both show two very different patterns within each region. The important point of this finding is that it shows that regions are not necessarily integrated into global service markets through just one pattern of linkages, as might be assumed. Finally, all the regional patterns show transregional linkages, but this is most evident in the Anglo-American service field. Linking European cities with cities created world-wide by settlers of (North) European origin centuries ago, this most global service field is the one based historically on migrations that happened centuries before globalization was even thought of. This illustrates the variety of the processes that have operated to globalize financial and business services.

The constant that emerges from studying the geography of services at the global scale is the hierarchical tendency within the network of world cities. The chief 'articulator' cities – Frankfurt/Munich, Chicago, Tokyo/Bangkok, New York, New York/Washington, DC, and London – are quite predictable, given reasonable knowledge of the globalization of services. And, of course, like the hinterworlds described in the previous chapter, the service fields articulated by these cities are global. With electronic communication, services can be organized in very different ways that do not need the simple contiguity of service-centre and service-field as traditionally occurs in analyses of services. This is clearly illustrated in what is the most interesting geography of the analysis. New York's extramural service field in advertising reinforces previous findings that global advertising requires local components even while being organized through a service hub that happens to be on another continent.

Comparison with the configuration of the global media network

The ultimate purpose of any quantitative measurement and analysis is comparison. If equivalent tools are used, then different segments of reality can be placed side by side and their similarities and differences evaluated. It is hoped that such matching exercises provide new insights into one or other of the items under investigation. In this final section I employ such a comparative strategy to see whether the configuration I have uncovered above is replicated to any degree when a similar principal components analysis is carried out on the media conglomerates data introduced for the comparative connectivities section of Chapter 5.

In this second principal components analysis there are thirty-three media conglomerates and I focus upon those cities that have at least five media firms present. This creates a 33 (firms) × 104 (cities) data matrix for analysis. In this case eight components are extracted and interpreted, details of which can be found in Kratke and Taylor (forthcoming). The purpose here is not to present these results for a different economic sector in the whole but to selectively draw out results that are interesting in relation to the corporate services analysis. This involves the first five of the eight new components; the three not discussed here are two particular patterns that relate to Italian cities and Scandinavian cities, plus a rather general media field without an articulator city.

I have five selected media fields to compare to six business service fields. In this exercise I find that one of the media fields has no service field equivalent and that two service fields have no media field equivalents. None of the three unpaired fields is a surprise. Paris articulates a media field that reflects its traditional predominance in this industry, with particular reference to worldwide projection of French language products; for instance, Montreal is an important primary field city in the analysis. Although Paris is an important global service centre, it was previously found that Paris did not articulate its own service field. The two service fields that have no media equivalents are the Tokyo/Bangkok-articulated Pacific Asian-biased field and the Chicago-articulated United States-biased field. These are hardly surprising given Tokyo, Bangkok and Chicago's relative unimportance as global media cities, as illustrated previously in Table 4.6. This leaves four matching pairs that exhibit some interesting similarities.

New York/Los Angeles United States-biased media field (media component I)

This media field (Figure 6.7) has similarities with the New York/Washington law component. Here New York has a different partner but the US bias is similar, as is the inclusion of important cities in Western Europe and Pacific Asia in the media field. This is less clear in this case because of the lack of any primary field cities. What both components primarily show is a New York-led dual articulation in economic sectors in which the United States dominates globally.

Munich/Berlin European-biased media field (media component II)

This media field (Figure 6.8) has similarities with the Frankfurt/Munich pan-European service field linking law and banking. Here Munich has a different partner, with Berlin, the new state and media capital, replacing Frankfurt, the international financial centre. In this case the geography is more north–south across Europe than east–west. Nevertheless, it is found once again that a city articulation from within Germany operates as the European heartland.

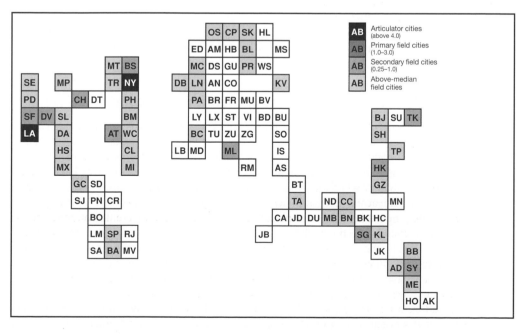

Figure 6.7 New York/Los Angeles United States media field (from Kratke and Taylor forthcoming). (For city codes, see Figure 4.12.)

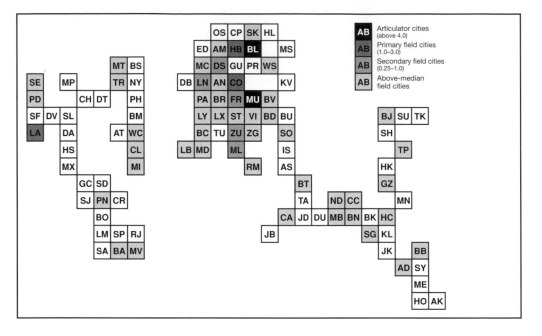

Figure 6.8 Munich/Berlin's European media field (from Kratke and Taylor forthcoming). (For city codes, see Figure 4.12.)

London-articulated general media field (media component III)

As with the Anglo-American accountancy component, London articulates a quite diffuse media field (Figure 6.9). Although not as pronounced as in the service field, this media field also features 'Old Commonwealth' cities and US cities. Both fields confirm London as a strategic hub of the world city network as well as a highly connected one.

New York-articulated European media field (media component V)

It is truly remarkable that New York appears once again as an extramural articulator to European cities (Figure 6.10), as previously shown for the European 'minor primate' advertising component. In this case the European cities are different, more heartland than minor primate, but the peculiar transnational arrangement is essentially the same. It is comparisons like these that make comparisons worthwhile.

What is to be learned from this exercise in comparisons? There are five important lessons. First, the complexity of globalization as a multifarious bundle of processes is again demonstrated. Second, once research goes beyond the particular focus on advanced producer services, there are other patterns that emerge, in this case with Paris emerging as a strategic articulator. Third, New York and London are confirmed as 'all-round world cities' – they have their own articulated fields in both analyses. Fourth, of the two leading world cities, New York, while slightly behind London in connectivities, appears to over-take London in importance once I begin looking at strategic patterning. Fifth, German cities appear strategically more important in the world city network than their levels of network connectivity would suggest. The last two points are particularly important since they show how the comparative multivariate analyses are not just recycling inferences

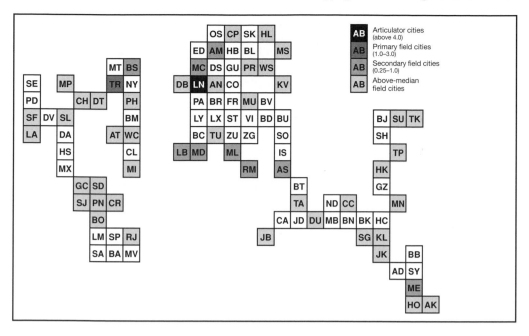

Figure 6.9 London's general media field (from Kratke and Taylor forthcoming). (For city codes, see Figure 4.12.)

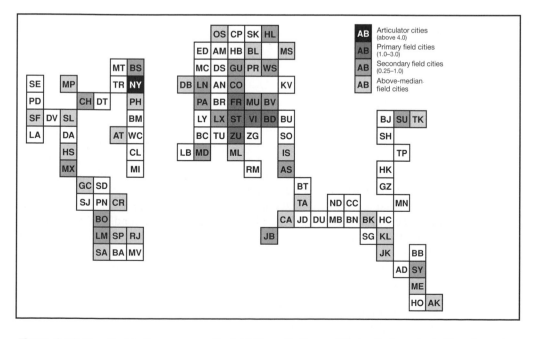

Figure 6.10 New York's European media field (from Kratke and Taylor forthcoming). (For city codes, see Figure 4.12.)

that can be drawn from simpler previous analyses in Part II. For instance, these comparative multivariate configurations are picking up the possibility of a specific state-effect impinging on transnational processes – cities in both the United States, with the largest domestic market in the world, and Germany, with the largest domestic market in the European Union, are network hubs that appear to boost their importance beyond their network connectivity levels.

6 Mappings of cities in globalization

> New economic geographies are a fact, though it would be better if we had an
> improved sense of their dimensions.
>
> (Cox 1997: 17)

In the mapping of services, most of the 123 cities in the analysis were barely featured.
The principal components analyses of 100 service firms' office distributions highlighted
the important cities for particular services, a few articulator cities and a few more
primary field cities, but with the great majority of cities merely appearing in some positive fields and not in others. The reason for this is quite simple: it was the variation
among service firms that was being analysed and not the variation among the cities.
In this chapter I focus on the latter. I use the same interlocking model, the same large
100 × 123 service values matrix and, for the most part, the same multivariate technique,
but I turn the analysis around to bring cities to the fore. Hence below I produce mappings
of cities rather than of services.

   In addition, I will be a little more sophisticated in the analyses in this chapter. In the
first section the same basic principal components analysis method used for delineating
the global configurations of services is applied again to cities. The resulting global
configuration of cities is interpreted as the 'prime structure' of the data, implying, of
course, that there are other structures. In the second section I use principal components
analysis in an exploratory manner whereby several further structures are identified. This
is about as far as I can take principal components analysis and I turn to a new technique, fuzzy set analysis, to take the analysis forward. This technique includes much of
the flexibility of experimental principal components analysis and can be applied to larger,
sparser matrices. In the third section I present an analysis including 234 cities and
produce a comprehensive set of results identifying twenty-two global urban arenas. This
is the final configuration derived directly from the data matrix of service values. In the
final section there is a first summary of such configurations.

Common profiles of city services

In the matrix of service values every column shows the global locational strategy of a
firm through the distribution of its offices across cities. This is the concern of the
previous chapter. Here I focus on the rows of the matrix. Each row shows the global
services mix or profile of a city as the distribution of its offices across firms.

Turning principal components analysis around

Every data matrix is the result of a data production exercise that involves measuring a
variable across a set of objects. By convention, the variables are arrayed as the columns

of the matrix. In the case of the matrix of service values, the variables are identified with the locational strategies of firms: as the description of data production in Chapter 3 made clear, the service values are the variables created firm by firm. Thus the ensuing matrix has each column as a firm strategy-variable. This leaves the cities as the objects upon which the measurements are arrayed as rows of the matrix of service values. In statistical analyses it is the variables that are usually analysed; it is their variances and co-variances that are at the heart of statistical techniques. This means that the principal components analysis using firm strategy-variables described in the previous chapter is the standard way in which the technique has been applied. But it is not the only way.

In the factor analytic family of techniques the derivation of new dimensions from variables is called R-mode analysis. The techniques were originally developed for just this task and it continues to be their typical mode of analysis. However, it is just as statistically feasible to analyse from the perspective of the objects. In a principal components analysis this involves creating components out of the correlations between rows – that is to say, focusing on similarities between objects. This is called a Q-mode analysis. A principal components analysis in Q mode produces component clusters of like objects (cities) instead of component clusters of like variables (firms) as described in the previous chapter. Importantly, the same property of parsimony obtains irrespective of analysis mode. Thus the idea is to reduce, say, ninety objects to just six 'composite objects' that account for a reasonably large amount of variation across all objects. In simplest terms, whereas parsimony was achieved by the R-mode analysis seeking out common patterns of similarity between service firm strategies, it is Q-mode analyses seeking out patterns of similarity between city service mixes that are described in what follows.

The main practical effect of 'turning around' principal components analysis is that it is now the cities that will load on (correlate with) the components that are extracted, and the component scores will be produced for the firms. Thus I am dealing with a multidimensional city space, or more accurately a service values space, this time defined through axes of cities rather than of firms. This is crucial because, unlike the service space defined by firms, in this city space all cities are treated equally. The variances of all city service profiles are standardized to 1, which means they each exist in the space as vectors of the same length, unity. In the analysis, this variance will be concentrated in just a few dimensions as before, but now all cities are directly part of the creation of common dimensions of variance. Thus the representation of the cities in this analysis covers much more than just the leading cities.

The primary configuration of cities

In the previous chapter it is pointed out that the choice of number of components is essentially a judgement call by the researcher to produce an interesting set of common patterns. I keep to this position in this section and have selected a five-component solution for initial discussion. I consider the five components to provide the best single description of common city profiles in the data and thus I interpret this solution as revealing the 'primary structure' of the data. This judgement is based on two considerations. First, this solution produces a reasonably balanced pattern of components in numerical terms (Table 7.1). Each component has at least five cities with high loadings (over 0.6) and each has a sizeable total of cities in its cluster (the two smallest components, IV and V, each have eighteen cities). Second, the five-component solution produces a structure that is very easily interpreted. Hence, as well as the numerical balance, these results suggest a clear geographical logic to the way clusters of city profiles are distributed across the world.

Table 7.1 *Primary configuration: city loadings*

I Outer cities	II US cities	III Pacific Asian cities	IV Euro-German cities	V Old Commw'th cities
784 Tel Aviv	769 St Louis	740 Taipei	782 Berlin	716 Perth
767 Sofia	703 Cleveland	726 Tokyo	768 Munich	715 Adelaide
753 Kuwait		725 Bangkok	703 Hamburg	
730 Helsinki		703 Jakarta		
730 Quito				
724 Beirut				
696 Casablanca	680 Dallas	664 Beijing	697 Cologne	687 Brisbane
681 Athens	664 Kansas City	658 Manila	660 Stuttgart	657 Hamilton (BD)
670 Nairobi	650 Pittsburgh	633 Seoul		616 Birmingham
666 Montevideo	634 Portland	630 Kuala Lumpur		
664 Jeddah	633 Atlanta	607 Hong Kong		
660 Bucharest	631 Seattle			
650 Indianapolis	623 Charlotte			
645 Cairo	622 Denver			
642 Lagos	620 Detroit			
629 Panama	607 Philadelphia			
624 Lima				
608 Vienna				
599 Dubai	560 Boston	598 Guangzhou	593 Frankfurt	547 Manchester
595 Copenhagen	557 San Diego	593 Shanghai	569 Paris	504 Nassau
595 Oslo	524 Washington	560 Ho Chi Minh	530 Budapest	501 Vancouver
592 Zagreb	524 Minneapolis	516 Istanbul	530 Düsseldorf	501 Nicosia
590 Karachi	502 San Francisco	511 Mumbai	519 Warsaw	
586 Chennai	500 Houston	500 Singapore	511 Milan	
584 Bangalore			508 Luxembourg	
572 Istanbul				
570 Lisbon				
553 Bratislava				
535 Kiev				
534 Nicosia				
533 Calcutta				
495 Riyadh	499 Melbourne	455 São Paulo	482 Antwerp	457 Abu Dhabi
492 Prague	473 Los Angeles	443 Caracas	460 Prague	453 Montreal
468 Auckland	462 Vancouver	416 New Delhi	452 Rome	442 Auckland
461 Moscow	437 Chicago	405 Santiago	437 Lyons	441 Calgary
457 Johannesbourg	425 Miami		433 Amsterdam	426 London
452 Cape Town	410 Montreal		402 Moscow	423 Dubai
448 Manila	409 Toronto			410 Port Louis
446 Budapest				408 Dublin
427 Mumbai				402 Wellington
424 Warsaw				
421 Port Louis				
418 Santiago				

The results of the principal components analysis will be described by interpreting the component loadings of cities (Table 7.1 and Figure 7.1) and the component scores of firms (Table 7.2) together. For the latter, I list just those component scores above 1.5. Table 7.2 provides an immediate contrast with the earlier R-mode analysis component scores (Figures 6.1–6.6). The fact that there is no component score above 3, the level at which I identified articulator cities in the R-mode analysis, is instructive. It means the distributions of service values as columns are very different from their distribution considered by rows. The result is that there are no dominant firms in the city components – 'articulator firms', perhaps – in the way that there are dominant cities that appear in firms' locational strategies. There is a second immediate result from initial perusal of the results, this time gained from the presentations of the loadings: membership of a component cluster in Table 7.1 is set the same as in the R-mode analysis (Box 6.1) at 0.4 and above. Again contrasting with the previous R-mode analysis, a wide range of cities is featured here as promised in proposing this Q-mode analysis. To get a clearer view of the spatial patterns, a threshold of 0.5 is used for inclusion in the clusters, as shown in Figure 7.1. Here it can be seen that the results divide geographically into two types: regional clusters of cities and inter-regional clusters of cities. I take each type in turn.

Regional city clusters

The regions identified are the globalization triad that I have referred to previously as the prime globalization arenas. The most straightforward regional result is that for component III describing Pacific Asian cities. The top twelve ranked cities (Table 7.1) are from the region, leaving out just Singapore (ranked fifteenth) as the most untypical of the region's

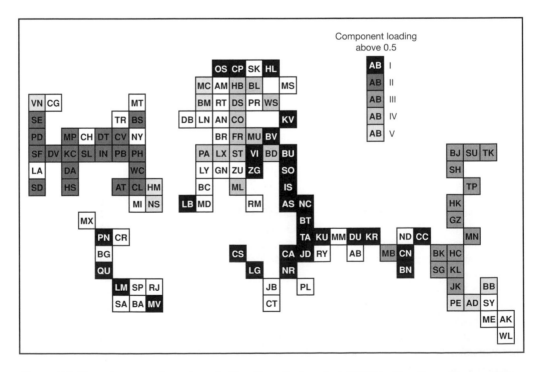

Figure 7.1 The primary configuration of cities (from Taylor *et al.* 2002b). (For city codes, see Figure 4.1.)

Table 7.2 Primary configuration: firm component scores

	I Outer cities	II US cities	III Pacific Asian cities	IV Euro-German cities	V Old Commonwealth cities
2.0 to 2.9	Ogilvy & Mather McCann-Erickson J Walter Thompson FCB BDO International IBM	Moores Rowland Chubb Group Towers Perrin KPMG **CREDIT SUISSE FB**	**SANWA** **DAI-ICHI KANGYO** **CITIGROUP** **HSBC**	**DEUTSCHE BANK**	CGNU Horwath International PKF International HLB International Royal & Sun Alliance **FUJI BANK** **CITIGROUP**
1.5 to 1.9	Deloitte Touche Toh. D'Arcy Masius B & B Young & Rubicam KPMG PricewaterhouseCoopers BBDO Worldwide Grant Thornton Int. Ernst & Young	Reliance Group Hold. **CHASE H & Q** Andersen Consulting J. Walter Thompson Watson Wyatt Arthur Andersen TMP Worldwide RSM International	**FUJI BANK** Young & Rubicam **BNP PARIBAS** **J P MORGAN** **BARCLAYS**	**COMMERZBANK** **BAYERISCHE LG** HLB International RSM International McKinsey & Co. **WESTLB** AGN International **ABN-AMRO** Allianz Group Winterthur **Linklaters Alliance** **DRESDNER BANK** Fudicial Int.	IGAF RSM International Lloyd's AGN Int.

Key: Accountancy, *Advertising*, **BANKING/FINANCE**, *Insurance*, **Law**, *Management Consultancy* cities. The service mix of these cities is dominated by banks (Table 7.2), indicating that this regional component is the converse of the banking component in the earlier R-mode analysis that showed a Pacific Asian emphasis in locational strategy (Figure 6.1).

The other two regional components are not simple R-mode reversals. The European cluster (component IV) is dominated by German cities (Table 7.1), but a quick glance at the component scores (Table 7.2) shows these cities are not clustered on the basis of legal services within their service profiles. Banks feature strongly but with accountancy firms featuring after banks. This difference is reflected in the geography by the omission of London from this city-cluster. However, the east–west dimension of the R-mode pan-European strategy (Figure 6.2) is maintained, with the representation of East European cities being greater than for Mediterranean and Scandinavian cities. The omission of Brussels shows once again that the European patterns I am finding here do not correspond at all with the EU and its common economic space.

The final regional cluster (component II) is a very compact one covering the United States: all the cities (18) with loadings over 0.5 are from this one country (Table 7.1). Los Angeles, Chicago and Miami also feature with smaller loadings, but New York is conspicuous by its absence. This shows clearly that the relatively less important cities can, and often do, define city-clusters in Q-mode analysis. New York, presumably like London, has a more 'globally orientated' rather than regional mix of firms. That is to say, these cities' mix of services is less dominated by firms with particular regional emphases in their locational strategies. In general, moving down the list of US cities in Table 7.1 leads to more externally orientated cities, culminating in the three cities noted above with loadings between 0.4 and 0.5. The scores for this component are also interesting: all service sectors are featured except law. Five sectors being represented shows US cities to be relatively 'well-rounded' global service centres; the lack of law reflects this service's relative unimportance for defining city clusters in general.

Inter-regional city clusters

Outside the globalization triad there is no region that constitutes a cluster of cities, rather there are two large transcontinental assemblages of cities. One of these clusters (component V) has a very specific interpretation: 'Old Commonwealth' cities (Table 7.1). This suggests an affinity with the accountancy component in the R-mode analysis (Table 6.1), which is confirmed in Table 7.2 wherein half the firms scoring on this component are accountancy firms. However, although in some ways a reversal of the R-mode component, this city cluster shows how differently the two modes of analysis treat cities. Here London, the articulator city in the R-mode analysis, is loaded quite low, and it is relatively unimportant Australian cities that have the high loadings (Table 7.1).

By far the largest number of cities (43) load on component I (Table 7.1), which I term simply 'outer cities'. 'Outer' is used here at two levels. In Europe the component encompasses cities on the edge of the continent in the north (Scandinavian cities) and the south (Lisbon, Athens). This very same pattern has been found in a different, earlier analysis focusing only on European cities (Taylor and Hoyler 2000). Beyond Europe, regions outside the globalization triad are featured, notably Middle Eastern and Latin American cities but also including cities from South Asia and sub-Saharan Africa. The service sector dominating the component scores is advertising (Table 7.2). This reflects the large number of capital/media cities loading on this component in Table 7.1 and relates to the advertising component in the R-mode analysis (Box 6.1). Firms from the 'ubiquitous sector', accountancy, also feature prominently in the component scores, which is consistent with the fact that none of the more important world cities loads on this component. Indeed, it seems to represent a worldwide swathe of what have been called 'wannabe world cities' (Short *et al.* 2000) outside the main globalization arenas. This component is termed 'outer cities' because its members are in the world city network but beyond its densest connections.

Let me reprise what a clustering of cities into a principal component means in terms of these data. The components are derived from the pattern of correlations between cities. Thus two cities loading high on the same component will share similarities in the service firms each house and in the 'service values' those firms give to each city. Therefore, components are clusters of cities with similar profiles of service provision across firms. In addition, these clusters of cities can be interpreted as indicators of sub-networks based upon flows (of information, of data, of orders) within firms' office networks. Given that firms pursue 'seamless service' for their clients, intra-firm connections should be larger than inter-firm service connections. In short, principal components as clusters of cities can be interpreted as describing space-of-flow subdivisions within the world city network. With this in mind, and using the large outer component as a frame, Figure 7.2 is a schematic arrangement that summarizes the results. It shows four largely distinct city components with little overlap among them but with three of them overlapping with the Outer Cities component. The exception is the US cities cluster, which is separate from the Outer Cities and has just a little overlap with the Old Commonwealth cluster. This is a further indication of the separateness of US cities within the world city network.

Uncommon cities

Not all cities fit neatly into the five city components. There are two sorts of such 'uncommon cities'. First, there are the hybrid cities to be found in the overlaps just discussed. With their dual membership of clusters, these cities constitute a small set (14)

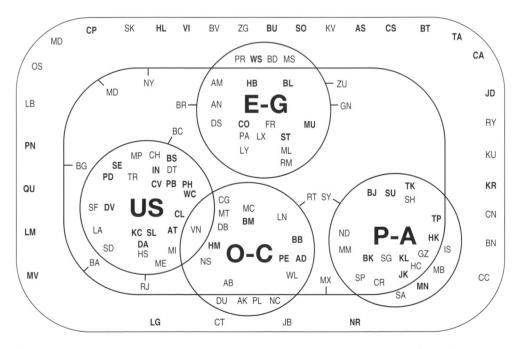

Figure 7.2 Schematic representation of the primary configuration (from Taylor *et al.* 2002b). (For city codes, see Figure 4.1.)

of interesting cities. Second, there are excluded cities, those that did not meet the threshold of loading at or above 0.4 for membership of any cluster. These are shown in Figure 7.2 attached to the cluster with which they have their highest loading. These twelve cities also demand our attention.

Table 7.3 shows the hybrid cities and emphasizes the role of the Outer City component in their formation. The clearest set of hybrids are the Eastern European former Comecon capital cities, a mini-region frontier between a major globalization arena and the world beyond. In contrast, the hybrid cities between Outer Cities and Pacific-Asian Cities are a little pot-pourri. The table identifies Manila as being different from other Pacific Asian cities; by loading on Outer Cities it is distinctively less core-like. If we look at the third overlap in Table 7.3 from another direction, the Australian and Canadian cities in the Old Commonwealth cluster are the ones that do not feature as outer hybrids, leaving cross-membership to Old Commonwealth cities in smaller economies.

In the one overlap not featuring Outer Cities (Table 7.3), the hybrid cities represent a relatively rare example where Canadian cities are combined with US cities in an analysis. In addition, in the initial loadings (Table 7.1), Toronto loads weakly on the US Cities cluster, further supporting the existence of this US–Canadian cities connection.

The twelve non-allocated cities, the excluded cities, are listed in Table 7.4 under the component on which they load highest. Some of the low loadings in Table 7.4 make sense – for instance, the three European Cities allocated to component IV, Bogotá listed under Outer Cities, and Sydney being linked to the Pacific Asian cluster of cities. But generally, this allocation to the 'closest' cluster tends to confirm the decision to denote them as excluded from this five-component patterning of cities. Other excluded European cities are to be found with highest loadings on each of the non-European

Table 7.3 *Primary configuration: hybrid cities*

	I – Outer	III – Pacific Asia
Istanbul	0.57	0.52
Mumbai	0.43	0.51
Manila	0.45	0.66
Santiago	0.42	0.41
		IV – Euro-German
Prague	0.49	0.46
Moscow	0.46	0.40
Budapest	0.45	0.53
Warsaw	0.42	0.52
		V = Old Comm.
Dubai	0.60	0.42
Nicosia	0.53	0.50
Auckland	0.47	0.44
Port Louis	0.42	0.41
	II – USA	
Vancouver	0.46	0.50
Montreal	0.41	0.45

Table 7.4 *Highest loadings of unallocated cities to five components*

I Outer cities	II US cities	III Pacific Asia cities	IV Euro-German cities	V Old C'wealth cities
0.39 Bogotá	0.38 Rio de Janeiro	0.37 Sydney	0.37 Geneva	0.30 Rotterdam
0.37 Madrid	0.38 Barcelona		0.36 Brussels	
0.35 Buenos Aires			0.23 Zurich	
0.30 New York				
0.29 Mexico City				

components. Even more instructive is the fact that the Outer Cities cluster picks up major world cities, including New York, using these low levels of loading. This suggests that the 0.4 threshold used to define clusters in Table 7.1 is a reasonable cut-off point before the onset of relatively arbitrary cluster allocations based upon low loadings that distort the results.

It is important to understand that these 'excluded cities' are excluded only at the level of the five-component solution. They do not fit into the primary structure of world cities as defined here but that does not mean that they are distinctive in general. With additional components it might well be that some or all of the twelve excluded cities in Table 7.4 would become allocated to new clusters that are yet to be defined. To see whether this is the case requires further empirical analysis.

Searching for further common city profiles

Calling the results of the Q-mode principal components analysis a 'primary structure' has left the door open to ideas about other structures in the data. In the previous chapter I introduced principal components analysis as a reordering of spatial dimensions creating the same number of components as original variables. Selection of a small number of components wherein the original variance is concentrated is thus a subjective decision. Indeed, in selecting a solution to present from an analysis, it is commonplace to consider other solutions and reject them. How else is a researcher to know that the selected solution is better than others? For instance, in selecting the five-component solution above, a six-solution result was inspected and found wanting: the sixth component included only ten cities with loadings above 0.4, only three of which were above 0.5, with none reaching 0.6. Component VI was thus so much 'weaker' than the first five components that it was deemed to contribute far less to the analysis. Thus it was decided that increasing the number of components to six provided a poorer, less clear-cut set of results; hence the decision to stay with just five components.

This dismissal of unwanted results is the usual way the question of how many components is tackled. But it is not the only way: principal components analysis can be seen as being very flexible and, instead of homing in on one solution, can open up deliberation to the many solutions. For that reason, principal components analysis makes an ideal exploratory statistical tool.

Beyond parsimony: principal components analysis as a tool for data exploration

Exploratory data analysis is usually limited to univariate or bivariate situations where patterns of data can be easily presented. Obviously, such an approach is not directly possible in a large multivariate context. Simple displays of the data remain an aim but their production has to be indirect following a deal of data manipulation.

Parsimonious description of large and complex sets of data is the main reason for using principal components analysis and I have employed this argument to justify use of the technique above. A good indication that this has become a standard argument for using a factor analysis is that in the standard software package for social science statistical analyses (SPSS), the family of factor analytic techniques is reached by clicking on the 'Data Reduction' button. However, in the classic social science text on factor analysis by Rummel (1970), a much wider range of uses is discussed. Rummel identifies ten different 'design goals' for these techniques (p. 182), one of which is to 'explore'. This is the approach I discuss here, but, alas, exploratory analysis is not followed up at all in his otherwise comprehensive text. I attempt an initial rectification of this oversight here.

Most statistical analyses, whether multivariate or not, are designed to provide an answer to a question. Answers range from averages (of frequency distributions) to gradients (of scatter plots) to clustering hierarchies (from data matrices), and in every case the research is carefully arranged to provide what is usually termed a 'best fit' answer to a research question. It is this way of thinking that I have employed thus far in the principal components analyses. Now I move away from the idea of finding a definitive answer and develop a different sort of research. The factor analytic family of techniques has always been vulnerable to criticism from those looking to find the 'one and only answer' in their data. Researchers carrying out factor analyses find themselves having to make more 'subjective' decisions than is normal for a quantitative analysis. This is because the method involves several 'sources of uncertainty' (Rummel 1970: 349). By far the most prominent of these is this question of the number of factors/components to extract and rotate. Choosing different numbers of factors produces different results – alternative answers to the same question – and there is no agreed way to decide how many is required. Notice that for the selections of number of components above, I had no recourse to the usual statistical criterion of 'best fit'; rather, I fell back upon the subjective criterion of interpretability. Rummel (ibid.) translates this uncertainty into 'the Factor Number Problem' and devotes a whole chapter to deciding 'the best number of factors' (ibid.: 351, 367). Unable to provide a solution to the problem, he concludes by admitting that the end result of his discussion has been to make the reader feel 'more confused' (ibid.: 367) than before! In fact, the problem is intractable unless the research design eshews single-solution thinking.

I take a different tack: rather than uncertainty being a problem, I look at it positively. Creating many alternative results provides a means for exploring a set of data. Instead of searching for the 'best number', I consider a wide range of possible numbers of factors. This 'multiple-number' design allows for comparison of results over a range of levels of 'data reduction'. In this way I uncover different sources of independent variation at different levels, but I also see the similarities in patterns over different levels. Such similarities indicate the robustness of some component patterns. Of course, this approach does not mean that all solutions are treated as being of equal relevance: exploration is all about evaluating the salience of different results using different numbers of factors. I know of no use of factor analysis in this exploratory manner, so the analyses below are presented tentatively as a methodological innovation for understanding large data matrices.

The actual research design consists of thirteen principal components analyses of the 100 × 123 city–firm data matrix of which the primary structure solution previously described is but one. The number of components extracted ranges from 2 to 14 wherein every solution has all components with at least one loading above 0.4. All these results are based upon varimax rotation to provide distinctive patterns of variation. The analytical route through the matrix is from most to least data reduction.

First cut: a basic global division of cities

The simplest principal components analysis consists of rotating two components to find the most basic sources of variation in the data. This is to dichotomize the data into two broad clusters of cities with distinctively different profiles of firms. In fact, because I allocate cities to components only where they have a loading of above 0.4, this produces a third group of unallocated cities. This exercise produces a first component of sixty-six cities and a second component of thirty-eight cities, and leaves nineteen cities unallocated.

The specific results of the two-component analysis are shown in Table 7.5 and Figure 7.3. It is instructive to begin with the unallocated cities. Although they are a rather motley crew covering all parts of the world, there is one notable feature of this group: it encompasses many major world cities. As well as London, New York and Tokyo, this group of nineteen cities includes Chicago, Frankfurt, Singapore, Sydney and Zurich. Even where major cities are allocated to the components, these are in the lowest category of loading (Hong Kong, Paris and Brussels in component I; Toronto, Los Angeles and Washington in component II). In complete contrast, the upper reaches

Table 7.5 *Simple division: city loadings*

	I – Outer wannabes	*II – Inner wannabes*
0.7+	Istanbul, Athens, Cairo, Montevideo, Sofia, Beirut, Prague	St Louis, Indianapolis
0.60–0.69	Dubai, Bucharest, Mumbai, Karachi, Tel Aviv, Budapest, Casablanca, Nairobi, Manila, Zagreb, Warsaw, Lisbon, Santiago, Quito, Moscow, Taipei	Charlotte, Kansas City, Atlanta, Seattle, Vancouver, Perth, Pittsburgh, Brisbane, Denver, Manchester, Adelaide
0.50–0.59	Panama City, Kuwait, Calcutta, Jakarta, Bangalore, Chennai, Caracas, Seoul, Kuala Lumpur, Lima, Vienna, Kiev, Johannesburg, Auckland*, Jeddah, Madrid, Amsterdam, Nicosia, Helsinki, Copenhagen, Dublin, Ho Chi Minh City	Portland, Houston, Philadelphia, Boston, Dallas, Minneapolis, Cleveland, Montreal, Melbourne, Birmingham, Cape Town, San Diego, Auckland, Barcelona, Calgary
0.40–0.49	Lagos, Milan, Port Louis, Hamburg, Bogotá, Hong Kong, Shanghai, Bratislava, Beijing, Buenos Aires, Guangzhou, Paris, Bangkok, Oslo, New Delhi, Geneva, Brussels, Stuttgart, Manama, Riyadh, São Paulo	San Francisco, Toronto, Detroit, Los Angeles, Miami, Lyons, Rome, Washington, Rio de Janeiro, Abu Dhabi, Rotterdam, Wellington, Hamilton (BD)

Note: Cities are ranked by loadings in each category (* indicates second highest loading for a city). Cities unallocated to two components: Antwerp, Berlin, Chicago, Cologne, Düsseldorf, Frankfurt, Hamilton (BD), London, Luxembourg, Mexico City, Munich, Nassau, New York, Singapore, Stockholm, Sydney, Tokyo, Wellington, Zurich

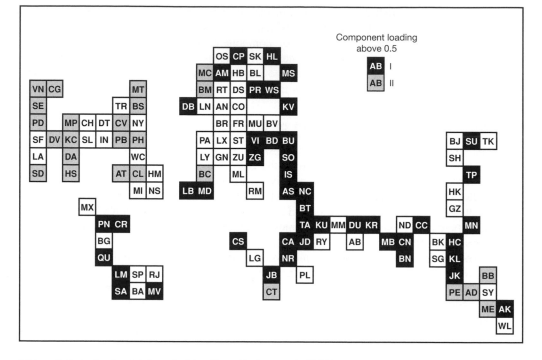

Figure 7.3 A simple division of cities (from Taylor *et al.* 2002b). (For city codes, see Figure 4.1)

of both components are replete with the relatively minor cities within the 123. I interpret both components as clusters of Short *et al.*'s (2000) 'wannabe world cities', important cities in their own right but outside the usual roster designated in world or global terms. Their generic name derives from the fact that such cities invariably have policies helping them strive for greater world city status.

Labelling these two 'wannabe' categories is quite straightforward. Component I is made up of cities from what used to be called the 'Third World' plus East European cities and some more peripherally located cities in Western Europe, notably in the far south (Mediterranean and Iberian cities) and far north (Scandinavian cities) – hence the designation 'Outer Wannabes'. Contrariwise, component II is dominated by relatively minor US cities plus the 'second cities' in West European countries (Manchester and Birmingham, both of which claim UK second city status, Barcelona, Lyons, Rome and Rotterdam) and second cities in selected associated countries (Montreal, Melbourne, Cape Town, Rio de Janeiro and Abu Dhabi). Clearly, these are 'Inner Wannabes'. This basic geographical dichotomy is clearly seen in Figure 7.3, where the first component straddles most of the world and the second component is concentrated in the 'old core', to use world-systems terminology. These are two distinctive policy worlds. For Outer Wannabes, rising up the ranks of world cities is primarily a 'development' issue, attracting global capital to become more central in the world city network. For Inner Wannabes, rising in world city status is about changing the nature of national city hierarchies in order to come out of the shadow of a dominant local world city. I have previously suggested that New York casts its 'world city shadow' over all other US cities (Beaverstock *et al.* 2000a), and clearly London plays a similar role in the United Kingdom.

Finally, it should be noted that in a usual principal component analysis research design, this two-component result would normally be overlooked, since in large, complex data sets it is highly unlikely that the 'best' solution would have only two dimensions. Usually such dichotomous results are found through using a hierarchical division analysis, a multivariate technique that does not have the flexibility of a series of principal components analyses. The former is strictly rigid in its creation of a hierarchy of clusters: once an object is allocated to an initial dichotomy of the data, it remains in that half of the data for all subsequent divisions down the hierarchy. In contrast, each Q-mode principal component solution of different numbers of component-clusters is independently created, and therefore an object (city) can move across the 'first cut' in subsequent reshuffling of the data. This is vital for exploring a data set because it allows a tracing of variable relations between components in different solutions. This can be illustrated by comparing the 'first cut' with the primary structure solution.

Table 7.6 shows a cross-tabulation of cities between the components in the two- and five-component solutions. There are two features that stand out. First, there are only two zero cells: no cities are members of both US Cities and Outer Wannabees or Pacific Asian Cities and Inner Wannabees. These apart, this means that cities in the five-component solution's clusters are coming from both clusters in the two-component solution as well as drawing from the unallocated cities. But the patterns of exchange are by no means even, and this brings us to the second outstanding feature. It is, of course, no coincidence that the term 'outer' is used to label components in both solutions: Outer Wannabees and Outer Cities have the largest overlap in Table 7.6. The former might be thought of as a sort of 'seed' for the latter. Similarly, 'seed' relations appear to link Pacific Asian Cities to Outer Wannabees and US Cities and Old Commonwealth Cities to Inner Wannabees. The Euro-German Cities are rather more mixed with respect to the two-component solution and take as many cities from the unallocated set as elsewhere. In a rigid hierarchical division it is unlikely that this cluster would be able to form.

New configurations

In this subsection I have chosen to focus initially upon the ten-component solution to illustrate further clusters of cities that I shall term secondary structures in the data (Table 7.7). This solution shows clearly three new city clusters and also illustrates the degeneration into very weak components as the number of components increases. In addition, it supports the previous identification of a primary structure, as a simple comparison between Table 7.1 and the first five components in Table 7.7 reveals. Both sets of five components are very similar and in Table 7.7 the importance of components immediately drops after component V. Although this shows the primary structure to be robust, the two sets of five components are not exactly the same. As well as a different order of extraction (IV and V are transposed between solutions), two of the prime factors alter

Table 7.6 *Cross-tabulation of primary configuration against simple division*

	I	II	III	IV	V	Not allocated	Totals
I	37	0	17	7	1	4	66
II	0	22	0	2	11	3	38
Not allocated	2	1	2	7	1	4	19
Totals	39	23	19	16	13	13	123

Table 7.7 A ten-component solution

	I Outer cities	II Minor US cities	III Pacific Asian cities	IV Old C'wealth cities	V German– E. Eur. cities	VI Larger Americas cities	VII Western European cities	VIII Indian cities	IX Minor Financ. centres	X Stockholm
0.7+	Tel Aviv Sofia Beirut Kuwait Quito Montevideo	St Louis Kansas C. Charlotte Cleveland Pittsburgh Indianap.	Bangkok Taipei	Perth Adelaide Brisbane	Munich Berlin Stuttgart					Stockholm
0.60 to 0.69	Casablanca Helsinki Athens Cairo Lima Nairobi Bucharest Lagos Panama Dubai Jeddah Zagreb	Denver Portland	Tokyo Beijing Jakarta Hong Kong Shanghai Seoul Manila	Birm'ham Manchester	Hamburg Düsseldforf Budapest		Milan Madrid	Chennai	Manama	
0.50 to 0.59	Istanbul Bratislava Oslo Vienna Copenhag. Karachi Kiev Nicosia Lisbon Bangalore Moscow Prague	Dallas Seattle Minneap. Philadel. San Diego Detroit Wash., DC Atlanta	Kuala L'pur Ho Chi M. C. Guangzhou	Vancouver Montreal Hamilton (BD) Calgary	Warsaw Cologne Frankfurt	Mexico C. Toronto Chicago Zurich Melbourne	Paris	Guangzhou	Luxemb'g	

Table 7.7 continued

	I *Outer* *cities*	*II* *Minor* *US* *cities*	*III* *Pacific* *Asian* *cities*	*IV* *Old* *C'wealth* *cities*	*V* *German–* *E. Eur.* *cities*	*VI* *Larger* *Americas* *cities*	*VII* *Western* *European* *cities*	*VIII* *Indian* *cities*	*IX* *Minor* *Financ.* *centres*	*X*
0.40 to 0.49	Chennai Riyadh Calcutta Budapest Manila Jo'burg Santiago Pt Louis Cape Tn Mumbai Auckland Warsaw Bogotá	Boston	Singapore Istanbul Mumbai Caracas	Auckland Cape Tn Abu Dhabi Wellington	Prague Moscow Luxemb'g Shanghai	São Paulo Sydney Boston Buenos A. San Fran.	Rome	New Delhi Calcutta Bangalore Seattle	Antwerp Pt Louis	Stockholm

slightly with extraction of secondary structures. This reflects the autonomous nature of different analyses in this exploratory strategy and is a reminder that the primary structure I have identified defines the most important, relatively robust, components, though it should not be interpreted as the 'correct' answer from the 'right' analysis. This will become clear as I interpret secondary structures by taking each new component in turn.

Component VI has no high loadings but nevertheless includes eleven cities in Table 7.7. Since the cities are relatively important world cities and seven are from North and South America, I label this component Larger Americas Cities. This necessitates relabelling (relative to the primary structure) the US component as Minor US Cities. Similarly, component VII has an altering effect on the European component from the primary structure. I label this next component Western European Cities, noting that it includes no German cities; these remain in the larger German-dominated European component. However, the latter now becomes a narrower component labelled German–East European Cities. Component VII is a rather minor component that I label Indian Cities (with Guangzhou as a hybrid case with Pacific Asian Cities). These cities come from the primary structure Outer Cities but this secondary structure is too weak to effect a substantial alteration to the original in this case. Component IX is a very minor dimension of common variation, incorporating just five cities that are each rather minor but with a banking speciality: my label is Minor Financial Centres. This interpretation is supported by cities with loadings just below the 0.4 threshold and not appearing in Table 7.7: Hamilton (Bermuda), Lyons and Geneva. Finally, Component X looks very insignificant, with just a single low loading, but again looking below the 0.4 threshold suggests that this Stockholm component might represent an incipient Scandinavian Cities cluster: Copenhagen, Oslo, Helsinki and Amsterdam (with its traditional Baltic links) are just below the threshold. However, having to search through relatively low loadings to make sense of the last two components does illustrate that these are very minor patterns of common variation; I consider just the initial three new components as representing secondary structures.

The three secondary structures are not necessarily best portrayed in the ten-component solution. Looking through other solutions, I have selected similar components that have a tighter structure in Table 7.8. For instance, the rather 'loose' (no high loadings) Americas component is more clearly specified as Major Latin American Cities in component VI of the six-component solution. With US cities not loading on this version, the

Table 7.8 Secondary structures

	VI with 6 components	VII with 7 components	VII with 9 components
	Major Latin American cities	Indian cities	Western European cities
0.6 to 0.69	Buenos Aires Caracas	Chennai	Milan Madrid
0.5 to 0.59	São Paulo Mexico City	Guangzhou New Delhi	Paris
0.4 to 0.49	Toronto Sydney Melbourne Madrid Milan Zurich Santiago Johannesburg	Calcutta Bangalore Detroit	Amsterdam Brussels Cologne

prime structure US Cities cluster label is retained. Note, however, that minor Latin American Cities remain in the Outer Cities cluster, suggesting a clear size differential among city service profiles of this region. Similarly, component VII of the seven-component solution provides a slightly more precise specification of Indian Cities, with New Delhi becoming more important. However, Mumbai still remains in the Pacific Asia cluster, suggesting another size differential between cities of a region, although this time it is the smaller cities that define the regional cluster. Finally, the Western European Cities are defined more clearly by component VII in the nine-component solution, which includes six cities including one German city, Cologne, suggesting that this city is less connected to East European cities than are other German cities.

In describing this research design I have used the metaphor of exploration. I think this has shown itself to be a good metaphor for the way it encompasses the methodological purpose. To explore is to venture into unknown territory where there is no room for rigid thinking; a flexible approach is a necessity. Because of the inherent 'uncertainties' in the application of the factor analytic family of techniques, their analyses can be adapted to a simple exploratory research design that is immensely flexible. Hence I have been able to illustrate how principal components analysis can be used as a tool for exploring a large, complex data matrix. Without searching for definitive answers, I have uncovered some of the basic patterns that constitute the data. The closest I come to a traditional analytical 'answer' is in the identification of the five-component solution as a 'primary structure' in the data. However, it should be pointed out that this was not an aim in embarking on the exploration. This primary structure was encountered *en route* through the data. In applications of this research design with other data there may well be no such primary structure. There are, of course, many multivariate techniques that can be employed to analyse large data matrices, but I think that employing principal components analysis as an exploratory tool is arguably the most fruitful way to begin an understanding. The results above are a justification for this assertion.

I do not produce neat findings. There are overlaps between clusters of cities, some cities are not allocated, and the content of clusters alters through different analyses. There is definitely no simple hierarchy of world cities. Nevertheless, I do finish up with a reasonable understanding of the structure of the data. This can be summarized as a primary pattern of five common sources of variation, three smaller secondary structures and two possible minor patterns, with the global cities of London and New York weakly represented across all structures. This is a new geography of globalization, as indicated by the configuration of world cities that was promised at the outset. It is offered as a rare research output: a sound empirical depiction of globalization, based as it is on a precise specification of the world city network and a careful measurement of the global strategies of 100 major service firms.

A larger mapping using fuzzy set analysis

In this section I take a new look at the original data with a view to incorporating more cities than the 123 used so far. Remember that the latter are from an initial roster of 315 cities, and therefore the scope exists for adding more cities to a multivariate analysis. The problem is, as pointed out in justifying selection of the 123, that as more less-well-connected cities are included, the matrix of service values becomes relatively sparse. In such a matrix there will be cities whose service profiles are dominated by zeros and therefore cannot be analysed meaningfully. However, the cut-off point does not have to be the 123rd most connected city. I have always regarded this as an arbitrary selection; here I consider a larger roster of cities.

Fuzzy set analysis is from a different family of multivariate techniques than principal components analysis: classification cluster analyses. These techniques use measures of similarity between objects to classify them into clusters (for details, see Derudder and Witlox 2002). Usually these clusters are formed as discrete sets of objects, each object having membership of just one cluster. In fuzzy set analysis each object has a level of membership in each cluster. Of course, in any reasonably structured data, it is expected that most objects have a high level of membership in just one or two clusters. Nevertheless, the emphasis remains upon 'hybrid' membership by objects, thus allowing more complex patterns of relations to be discerned. These membership levels are superficially like the loadings in the Q-mode principal components analysis – remember, every object loads on every component – but despite similarity in usage, they derive from totally different models. The key point here is that fuzzy clustering is less sensitive to sparse matrices and therefore I can explore the service values matrix from a new, and more comprehensive, perspective. I have selected all cities that have a site value of at least 20 (see equation 3.3). This yields a new set of 234 cities for analysis.

As with principal components analysis, different number of clusters can yield different salient results, and hence there is no firm theoretical basis for selecting the number of clusters from the classification analyses. Here, I will focus on the results for twenty-two clusters. This is a pragmatic choice after assessing several solutions. With this number of clusters I find a broad diversity in hierarchical and regional patterns in the world city network, one that provides for a particularly insightful interpretation.

Urban arenas

The twenty-two clusters in the fuzzy classification of 234 cities are intertwined in three different ways. First, there is a strong hierarchical dimension to the clusters: cities with similar levels of global network connectivity tend to be classified together. Second, there is a strong regional dimension to the clusters: cities from the same part of the world tend to be classified together. Third, there is a tendency for interaction between these two dimensions: clusters with low average connectivity tend to be more regionally restricted in membership. These geographical features mean that the results show more than clusters as segments of an abstract 'service space'; they have concrete expression as *urban arenas* in a geographical space that is the world city network.

This is an important interpretation because it indicates that cities are not creating and reacting to a simple process of globalization leading to an overarching world city hierarchy. There is a multifaceted geography of arenas through which cities operate as service centres for global capital. Hence as well as the commonplace notion that individual world cities represent critical local–global nexuses, there are also urban arenas that represent regional–global nexuses within contemporary globalization.

This new, complex global urban geography is shown in Table 7.9 and Figure 7.4. The table highlights the hierarchical tendency in the results, with arenas listed in terms of average global network connectivity for cluster members. These cluster connectivities are in turn used to denote five bands of arenas representing the hierarchical tendency around cluster A, which is by far the most important arena in terms of connectivity. The latter is called the 'Centre' of the bands for reasons that will become clear when Figure 7.4 is examined. The largest gap in connectivity is between cluster A, the Centre, and cluster B in Band I, but all the bands are identified using gaps in their levels of average connectivity. To give a feel for the structure and geography of the fuzzy classification, Table 7.9 shows also the size of each cluster, including overlapping cities, and the most typical city in each cluster.

Table 7.9 *Bands of arenas in the world city network*

Cluster/ Arena	Average connectivity	Band	No. Of members*	Typical city**
A	0.988	Centre	2 (0)	London
B	0.613	I	7 (2)	Frankfurt
C	0.574	I	3 (0)	Chicago
D	0.539	I	11 (2)	Amsterdam
E	0.438	II	8 (0)	Bangkok
F	0.401	II	5 (1)	Atlanta
G	0.384	II	4 (0)	Berlin
H	0.379	II	6 (0)	Warsaw
I	0.371	II	9 (5)	Istanbul
J	0.297	III	7 (4)	Caracas
K	0.297	III	12 (8)	Copenhagen
L	0.231	III	23 (6)	Adelaide
M	0.225	III	12 (2)	Calcutta
N	0.201	IV	9 (4)	Montevideo
O	0.193	IV	23 (0)	Baltimore
P	0.180	IV	16 (10)	La Paz
Q	0.179	IV	19 (12)	Kuwait
R	0.158	V	14 (5)	Leeds
S	0.157	V	8 (0)	Dresden
T	0.148	V	13 (3)	Lille
U	0.141	V	22 (8)	Accra
V	0.121	V	13 (3)	Osaka

Notes:

* Membership is defined as affiliation of 0.3 and above; figures in parentheses refer to hybrid cities with membership of other clusters

** Member with the highest affiliation

The regional tendency in the results is added to the hierarchical tendency in Figure 7.4: arenas are depicted in their respective bands around the Centre and in addition they are located in roughly their geographical position. The latter are articulated about a transatlantic centre of London and New York. Two member cities are shown for all arenas to aid in initial reading of the cartogram. In addition to the Centre arena, there are only three other arenas that have strong transregional membership, two in Band I, and one, perhaps surprisingly, in Band IV. This means that eighteen of the arenas have relatively clear-cut regional identities, thus showing the strength of the regional tendency in these results.

Further interrogation of the results requires a detailed look at the content of the arenas. For each cluster/arena I have searched for four sets of cities. First, the *cluster nucleus* is made up of those cities with membership levels above 0.7. Second, other *singular*

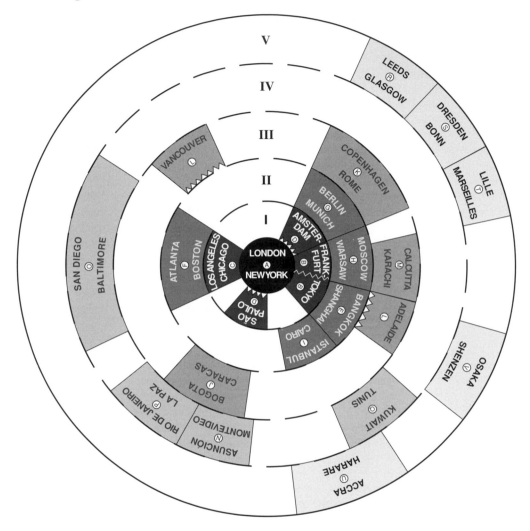

Figure 7.4 Urban arenas from fuzzy set analysis (from Derudder *et al.* forthcoming)

members are cities with membership levels between 0.3 and 0.7 and with no membership of another cluster. Third, *hybrid members* are the other members of the cluster that share membership with another cluster. Clusters without hybrid members I refer to as *distinctive*. Fourth, *near-isolates* are cities that are members of no clusters (they have no membership level as high as 0.3) but are designated through their highest membership level to a given cluster. As before, I begin by ordering the argument through hierarchical bands before focusing on the regional patterns.

The hierarchical tendency in urban arenas

Tables 7.10–7.14 show the cluster nuclei, singular members, hybrid members and near-isolates that compose the twenty-two urban arenas.

Table 7.10 *Centre and Band I arenas*

City type	A	B	C	D
Cluster nucleus	London	Frankfurt	Chicago	Amsterdam
	New York	Hong Kong	Los Angeles	Zurich
		Paris		Madrid
		Singapore		Milan
		Tokyo		São Paulo
				Mexico City
Singular members			San Francisco	Sydney
				Buenos Aires
				Toronto
Hybrid members		Brussels > D		Brussels > B
		Barcelona > D		Barcelona > B
Near-isolates				

Leading world cities: the Centre and Band I

The Centre is simple in the extreme. It consists of a two-city nucleus and nothing else. The Centre arena is 'Main Street, World-Economy'; by far the most important link in the world city network in terms of connectivities (Table 7.9), it is also a wholly distinctive dyad. This is the justification for terming the arena the Centre and so locating it as the pivot of the cartogram (Figure 7.4). Bringing together these two cities is a very important result of this analysis, one that the principal components analyses never achieved.

The Band I arenas are also relatively small and simple. This is especially the case with *Arena C*, which includes just the three US cities that rank below New York. It is a distinctive arena with no hybrids. The other two Band I arenas are cross-regional and link West European cities with cities in other parts of the world. *Arena B* links Paris and Frankfurt with the leading Pacific Asian cities. *Arena D* links other leading European cities with leading world cities outside the other two main globalization arenas (the United States and Pacific Asia) in European-settler regions, notably Latin America. These two arenas share Brussels and Barcelona as members. The distinction between the two arenas is the particular dominance of banking/finance services in the arena including Pacific Asia.

The Centre and Band I arenas define the twenty-one most important cities within this analysis of the world city network. They also suggest specific relations among these leading world cities. Beyond the Centre, it is the European cities that appear pivotal in linking to other regions; the leading US cities arena appears as relatively isolated in global service provision.

Major regional world cities: Bands II and III

The Band II and III arenas (Tables 7.11 and 7.12) are regional clusters of important world cities. There are three classic examples in Band II: *Arena E* is a distinctive cluster that includes all the important Pacific Asian cities not in Band I arenas; *Arena G* is a

Table 7.11 *Band II arenas*

City type	E	F	G	H	I
Cluster nucleus	Bangkok	Atlanta	Berlin	Warsaw	Istanbul
	Jakarta	Boston	Dusseldorf	Moscow	Dubai
	Kuala Lumpur	Dallas	Munich	Prague	
	Manila			St Petersbg	
	Seoul				
	Shanghai				
Singular members	Beijing	Washington	Hamburg	Budapest	Mumbai
	Taipei			Vienna	Cairo
Hybrid members		Miami > J			Dublin > I
					Lisbon > I
					Athens > I
					Amman > Q
					Beirut > Q
Near-isolates				Kiev	Geneva

distinctive cluster that includes all important German cities not in Band I arenas; and *Arena H* is a distinctive cluster that includes all the important East European world cities. The latter has a near-isolate, relatively unimportant city but appropriately located geographically for this arena, Kiev. Not quite distinctive but otherwise similar to the above arenas, *Arena F* includes the important US cities not in the Centre or Band I. In this case there is one hybrid member, Miami, which is linked into *Arena J* in Band III. The latter is the arena of leading Latin American cities that are not in Band I. Clearly, this analysis is picking up the regional articulation role of Miami between the United States and Latin America.

Arena J also links down the bands to less important Latin American arenas. This non-distinctive structure, sharing cities with other clusters, is typical of the other arenas in Bands II and III. *Arena I* brings together important Asian cities outside the Pacific Rim but also has links to a lower-band arena of Asian cities and to *Arena K*, which is a cluster of important 'outer' European cities. This odd combination, grouping Northern, South-Eastern and South-Western European cities, replicates previous findings in the principal components analysis above (Table 7.1) and previous analyses based only on European cities (Taylor and Hoyler 2000). *Arena M* is more distinctive than the others and is clearly a South Asian arena with just two hybrids. Finally, *Arena L* is a cross-regional cluster that covers the Old Commonwealth. This 'cultural' historical throwback arena again replicates previous findings above (Table 7.1). Membership covers Australian, Canadian, New Zealand and South African cities not found in Band I arenas. Note the dearth of British cities: only three appear as hybrids. They link to a particular British arena in Band V (Table 7.14).

With the exception of the latter unusual cluster, these arenas show that below the top echelons of the world city network, important cities tend to be very regional in the focus of their global service provision.

Table 7.12 *Band III arenas*

City type	J	K	L	M
Cluster nucleus	Caracas Bogotá	Copenhagen	Adelaide Brisbane Perth Vancouver Montreal	Calcutta Karachi Bangalore
Singular members	Medellín	Rome Stockholm Helsinki	Auckland Hamilton (BD) Cape Town Winnipeg Calgary Ottawa Christchurch Edmonton Johannesburg Melbourne Hobart	Islamabad Chennai Dhaka Riyadh Jeddah Lahore New Delhi
Hybrid members	Lima > P Santiago > P San José > N Miami > F	Dublin > I Lisbon > I Athens > I Riga > Q Vilnius > Q Tallinn > Q Sofia > Q Bratislava > Q	Canberra > P Monterrey > P Guadalajara > P Birmingham > R Manchester > R Southampton > R	Nairobi > Q Colombo > U
Near-isolates	Curitaba	Oslo	Durban Wellington Ruwi Manama	Ho Chi Minh C. Bucharest

Important cities on the edge: Bands IV and V

In Bands IV and V (Tables 7.13 and 7.14), I come to cities that were not covered in the previous principal components analysis. This does not mean, of course, that they are not involved in the same globalization processes as the cities dealt with above, but they are less intensively connected to the world city network. Given the conclusion concerning Band II and III arenas, I would expect the arenas in these two lower bands to be even more regional in their memberships. And this is indeed the case: there are three European arenas, two each from Asia and Latin America and one each from Africa and the United States. The last of these, *Arena O*, is distinctive and large and incorporates all remaining continental US cities if the near-isolates are included. The cluster of less important German cities, *Arena S*, is similarly distinctive but much smaller. It also

Table 7.13 *Band IV arenas*

City type	N	O	P	Q
Cluster nucleus	Montevideo Asunción	Baltimore Columbus Kansas City Richmond Charlotte Cincinatti New Orleans St Louis San Diego Indianapolis Portland	La Paz Quito	Kuwait Tel Aviv Tunis Dalian
Singular members	Port Louis Guayaquil Guatemala City	Sacramento Pittsburgh Tampa Phoenix Philadelphia Cleveland Minneapolis Buffalo Denver Hartford San Jose (CA) Detroit	Honolulu Porto Alegre Belo Horizonte Rio de Janeiro	Zagreb Casablanca Nicosia
Hybrid members	San José > J Panama > P S. Domingo > U S. Salvador > U		Kingston > U Managua > U P. of Spain > U Tegucigalpa > U Canberra > L Monterrey > L Guadalajara > L Lima > J Santiago > J Panama City > N	Bratislava > K Riga > K Vilnius > K Tallinn > K Sofia > K Nairobi > M Beirut > I Amman > I Tashkent > V Almaty > V Ankara > V Ljubljana > T
Near-isolates		Seattle Houston Rochester Las Vegas	Palo Alto Limassol Nassau	Abu Dhabi

Table 7.14 *Band V arenas*

City type	R	S	T	U	V
Cluster nucleus	Leeds	Dresden	Lille	Accra	Osaka
	Aberdeen	Bonn	Lyons	Dar es Sal.	Teheran
	Glasgow	Hanover	Marseilles	Gaborone	Shenzen
	Belfast	Nuremberg	Strasbourg	Kampala	Yangon
	Liverpool		Bordeaux	Lusaka	Yokohama
	Newcastle			Lagos	
				Harare	
Singular members	Bristol	Leipzig	Basle	Doula	Tianjin
	Edinburgh	Stuttgart	Lausanne	Brazilia	Guangzhou
	Nottingham	Cologne	Seville	Abidjan	Baku
		Utrecht	Bologna	Dakar	Labuan
			Berne	Doha	Nagoya
				Windhoek	
				Maputo	
Hybrid members	B'ham >L		Ljubljana >	Kingston > P	Tashkent > Q
	M'chester > L		Q, U	Managua > P	Almaty > Q
	South'ton> L			P. of Spain > P	Ankara > Q
			Bilbao > R	Teguc'pa > P	
	Bilbao > T		Valencia > R		
	Valencia > T			Colombo > M	
				Ljubljana >	
				Q, T	
				S. Dom'go >N	
				S. Sal'dor > N	
Near-isolates		Essen	Antwerp		Hanoi
		Rotterdam	Turin		
		Göteborg	Malmö		
		The Hague			
		Luxembourg			

includes some neighbouring European cities as singular members and near-isolates. Both the United Kingdom (*Arena R*) and France (*Arena T*) have their own urban arenas of less important cities, albeit less distinctive in nature than Arena S. The UK arena includes the Commonwealth arena hybrids; France includes other neighbouring European singular members, hybrids and near-isolates.

The Latin American and Asian clusters are much less clear-cut as geographical arenas. In particular, *Arenas N* and *P* both include less important Latin American cities from across the region with no obvious geographical division of the region. For instance, Central American and Caribbean cities are found in both clusters. In contrast, the Asian clusters, *Arenas Q* and *V*, have geographical concentrations in West and East Asia respectively. The former thus includes Middle Eastern cities not previously appearing in a cluster and the latter includes almost all the Pacific Asian cities not included in earlier clusters. Perhaps appropriately, they share Central Asian cities as hybrid members. Arena Q also shares less important outer European cities with Arena K. Finally, there is an African cluster: *Arena U*. All inter-tropical African cities belong to this arena except for Nairobi. Furthermore, all these African cities are part of the nucleus or are singular members; the hybrids of this arena are non-African. This reflects the lowly and relatively isolated position of inter-tropical African cities in the world city network.

Arena 'gaps': regional and national geographies

The findings are from a global urban analysis and this is how I have interpreted them above. However, the identification of urban arenas does point to some· interesting 'subglobal' conclusions. Specifically, I can comment upon how contemporary globalization seems to be impinging on long-established national and regional 'urban systems', the traditional concern of urban geographers studying relations between cities described in Chapter 1. Figure 7.4 is particularly informative in this respect.

First I can contrast the location of the arenas featuring cities in Europe's leading three economies. Whereas German cities are featured in three bands (I, II and V), British and French cities are concentrated in two bands (the Centre and V for the United Kingdom, I and V for France). This clearly shows the different national patterns of cities, with Germany's distinctive 'horizontal' city relations compared with the United Kingdom's and France's 'vertical' city relations. Although this is by no means a surprising result, it is relevant to the workings of contemporary globalization: whereas London and Paris cast an inhibiting 'shadow' over their compatriot cities, Frankfurt has no such effect on other German cities. Clearly, this is of vital importance to how each of these national economies relates to globalization processes.

US urban arenas have a similar pattern to German arenas except that they are represented in even more bands (the Centre, I, II and IV). This is perhaps to be expected given the large number of US cities within the data. However, this fact makes the lack of a US arena in Band III interesting. There appears to be a gap created in the globalization of US cities between the likes of relatively important cities such as Boston and less important cities such as Baltimore. This certainly implies policy incentives for cities in Arena O to try to 'move up' and create a new US arena in a higher band. More generally, the US arenas are typified by their high levels of distinctiveness. New York, as half of the Centre, and Miami as an important hybrid city linking to Latin America, are the only continental US cities to share arenas outside their own country. This relates to the sheer scale of the US economy and its long-developed, massive market in financial and business services, which provides less of an incentive for firms to 'go global' to the same degree as global service firms from other world regions. This highlighting of the ambiguous role of US cities in contemporary globalization is an important result of this research.

Finally, the most important result of this research is the light it shines on erstwhile 'Third World' cities in the world city network. The main point is that both Asian and Latin American urban arenas are quite well represented across all bands beyond the centre. In the case of Africa, its Arab cities and South African cities feature in relatively important Asian and Commonwealth arenas, but the inter-tropical cities all cluster in one Band V arena. The exception is Nairobi, which is a hybrid in two Asian clusters in Bands II and IV. Here there are the only signs of an authentic African world city emerging as more than just the NGO centre identified in Chapter 4.

Geographies of a nexus

As I indicate before the empirical analyses of the previous few chapters, my focus is upon the tight nexus between global service firms and world cities (Figure 3.1, p. 59). What have been produced in the two previous chapters are configurations of how that nexus is expressed as geographies of services and cities across the world. As indicated in Chapter 3, the other agencies in world city network formation – nation-states and service sectors – have been found to be important. As expected, much of the interpretation has focused upon these agencies for understanding particular patterns. The previous chapter was sector orientated; in this chapter I have had more to say about states. However, there are two features that dominate the geographies uncovered which I can term 'regional and hierarchical tendencies'.

The 'regional tendency' is for cities in the same part of the world to have similar mixes of global service firms. Thus although the latter have been selected to be global in scope, they display regional emphases that result in the geographies I have described above. Obviously, I have produced a snapshot of an ongoing process so that the firms may be globalizing but they are not necessarily 'fully globalized'. In this way the geographical origin of firms, usually indicated by their headquarters location, is immensely important to the resulting 'global geographies'. This is most clearly illustrated by the common separation of US cities in the various analyses. But this is not the only influence creating regionalities. In particular, different sectors have regional emphases in their globalizations. The most clear-cut example of this is the Pacific Asian city bias of banking/finance, and not just through Japanese banks. In short, globalization is a very regional process.

The 'hierarchical tendency' is for cities to be differentiated in terms of their levels of connectivity in the network. This was particularly evident in the large fuzzy set analysis as hierarchy and region intersected to create the urban arenas of cities with similar service mixes. Of course, these hierarchical processes are generated in the firms themselves, as shown in Chapter 4, but they carry through into an ordering of world cities in terms of high connectivities of world and regional headquarter cities. This is not exactly the same as Friedmann's original world city hierarchy but neither is it a simple reflection of a city network without hierarchical functions. The network I am dealing with is a very complex one involving hierarchical tendencies but with many other inter-city relations contributing to the configurations.

 **Part IV** **Suppositions**

8 From past to present: a metageographical argument

> Our historic time is defined by the transformation of our geographic space.
>
> (Castells 1999: 294)

> History does not repeat itself and city-states are unlikely to reappear. . . . [We are entering] a new historical interlude – whose stability and length we can only surmise – that may once more bring some political space to cities. This space may be limited, yet the room for manoeuvre is growing for cities.
>
> (Bagnasco and Le Gales 2000: 7)

Although this book is essentially an empirical work, in this final part I free myself from the strict evidential exigencies that have characterized the arguments thus far. Suppositions are speculations. I hope mine are not 'mere suppositions' but, rather, informed conjectures not wholly divorced from the evidence and models presented above. But I do reserve the right to go far beyond what I can safely infer from the global urban analyses. Effectively the latter have provided cross-sectional analyses about one segment of human activities at the start of the twenty-first century. In Part IV I develop an argument that this small time segment may represent a harbinger of a much broader shift in human organization. In this way I engage the debate concerning the meaning of contemporary globalization and provide a little new gloss on some existing ideas.

The existing ideas I develop are Wallerstein's (1979) world-systems analysis and Castells' (1996) network model of contemporary society. In each case the concepts and theories are used to draw out ideas that enable me to locate today's world city network in a general unfolding of social change in the modern world. I mesh these geohistorical interpretations with the contemporary cross-sectional analyses to create two sets of suppositions. First, I consider how the modern world got to this present, with suppositions that draw upon world-systems concepts. I present a particular geohistorical interpretation that culminates in the world city network in the historical tradition of defining the past to inform the present. Second, I look from the present world city network to speculate about where current trends might be leading. This is pure supposition; unlike in the case of historical change, I have no direct evidence to shape to my needs. I use the futurist tradition of excavating the past to explore future possibilities.

This chapter is about looking backwards to provide an interpretation of globalization through world city lenses and world-system concepts. The latter provides a geohistorical framework in the form of the long hegemony cycles introduced in Chapter 1 as part of the discussion of Braudel's world-cities. As defined by Wallerstein (1984), these cycles are usually differentiated in terms of the lead-state or hegemon – the Dutch Republic, Britain and the United States in turn. While not departing from this ordering and structure, I introduce a new way of looking at these cycles in terms of their metageographies. The concept of metageography is central to this interpretation and is the topic of the first section. Defined and explicated, metageographies are then used to interpret hegemonic cycles as different organizations and imaginations of world-space. Thus the second section describes three distinctive metageographies. The third section concentrates on the latest metageography characterized as 'embedded statism' which I describe in relation to cities and the study of cities. This leads directly to globalization and identification of a world city network and I ask the fundamental question of this chapter: do globalization and the world city network represent a transition to a new metageography? Thus while globalization most certainly does not mark the 'end of the state', it may signal the end of the state-centric metageography.

Metageography

Metageography is the term coined by Lewis and Wigen (1997) to describe the geographical structures through which people order their knowledge of the world. It is part of a society's taken-for-granted world. Rarely questioned as to its veracity or utility, a metageography constitutes an unexamined spatial discourse that provides the framework for thinking about the world across the whole gamut of human activities and interests. Lewis and Wigen name their book *The Myth of Continents* because this is a prime example of such accepted spatial dogma. The fact that Europe is a designated 'continent' has nothing to do with geography and all to do with history and culture. Geographically, Asia consists of a large central zone below the Arctic Ocean made up of great mountains and steppes from which six large peninsulas protrude: from east to west these are North-East Asia (Siberia), East Asia (China), South-East Asia ('Indo-China'), South Asia (India), West Asia (Arabia) and North-West Asia (Europe). Only the last projection is a continent; this privileging of a peninsula is the geographical origin of Eurocentrism. And yet the idea of a world divided into continents of which one is Europe is universally accepted as integral to the contemporary world-image, the largest scale of units in today's metageography.

As collective geographical imaginations, all societies will have distinctive metageographies. Well-known examples are the traditional Chinese view of the world centred on the 'Middle Kingdom', the Muslim division of the world into the 'House of Islam' and the 'House of War', the medieval Christian 'T and O' map with Jerusalem at the centre, and today's modern world normally portrayed as a map of 'nation-states'. These last are expected to fit into the continents to produce a nested pair of mosaics, and where this does not happen, in Russia and Turkey, it violates a geographical sense of order. All these world images constitute anchors that tie a society to both physical and metaphysical worlds; they provide the spatial configuration through which sense can be made of human social activities and ideals. As such, metageographies have been traditionally as much about cosmology as geography, at least in the way the latter is interpreted today.

But no metageography is eternal. I historicize the concept through adding the idea of a transition as a critical time of change between metageographies. Ultimately this will enable me to ask whether contemporary globalization constitutes a metageographical transition. In other words, is the current spatial anchor that helps make sense of the modern world – the map of nation-states – disintegrating in contemporary times? My route to answering this question is a geohistorical one that begins with an interpretation of the modern world-system in terms of metageographies and their transitions.

Metageographies of the modern world-system

The thesis developed here is that in the modern world-system, the balance between the physical and the metaphysical in the creation of metageographies has altered to the detriment of the latter. From the European discovery of the Americas through the Scientific Revolution and the secularization of the state to contemporary globalization, traditional cosmologies have been in retreat from an ongoing 'modernization' of knowledge. Following Berman (1988), I interpret modernity as indicating a state of perpetual social change which means that people living in modern society are in dire need of 'anchors', such as a metageography, to help stabilize their experiences. Since the modern world-system is the capitalist world-economy (Wallerstein 1979), it is the material basis of geography that becomes central to how the world is viewed. But it is precisely this material basis that is the motor of the incessant change. Thus it cannot be expected that a single metageography can survive through the unfolding of the capitalist world-economy. Hence the representative of the modern world in the list of examples of metageographies above – the map of nation-states – is not the metageography of the whole of the history of the modern world-system. In the modern world there has been more than one metageography.

Modernity has taken many forms over time and space since the emergence of the modern world-system in the 'long sixteenth century' (c. 1450–1650). I have argued elsewhere (Taylor 1999) that there have been three prime modernities in the history of the modern world-system. These modern constructions with their critical systemic repercussions are associated with the world-system's hegemonic cycles. The first hegemon, the United Provinces, was largely instrumental in creating a new mercantile modernity in the seventeenth century with navigation as the key practical knowledge. The second hegemon, the United Kingdom, was largely instrumental in creating an industrial modernity in the late eighteenth and nineteenth centuries with engineering as the key practical knowledge. The third hegemon, the United States, was largely instrumental in creating a consumer modernity with media/advertising as the key practical knowledge. I argue here that each of these modernities is associated with a distinctive metageography.

As spatial frameworks, modern metageographies are best identified by their geometries. The geographical metageography of mercantile modernity is a topological metageography of trade routes extending from the Philippines in the 'Far West' to the Moluccas (the Spice Islands) in the 'Far East'. For industrial modernity there is a centripetal metageography with the world seen as a single functional region serving a North Atlantic core. With consumer modernity there is a mosaic metageography of nation-states, national markets in which to ply wares. In each case there is a metageographical transition when the old is eroded, leaving a geographical opportunity for a new picture of the world to emerge.

Like all geographies, metageographies have three aspects: pattern, content and meaning. Hence after considering their metageographical transition I will briefly consider each metageography in turn in these terms.

The topological metageography of mercantile modernity

The metageographical transition that opened up the way to the first modern metageography is the archetypal example of the process. The traditional Christian cosmology simply could not contend with the European discovery of the Americas: adding a fourth continent made an imagination limited to three continents redundant (Zerubavel 1992). Of course, this 'discovery' also began the process that led to the famous 'triangular' Atlantic trade. At the same time, Europeans were developing routes through the Indian Ocean to the Pacific, avoiding Islam and the Muslims' control of Eurasian overland routes. The end result was that by the seventeenth century, for 'modern Europeans' the oceans represented a set of pathways linking together cities, ports, plantations, forts, mines and way-stations into a single trading system.

The topology of this metageography is shown in Figure 8.1, reflecting the global space of flows created by Europeans in the sixteenth and seventeenth centuries. There are five main European players, with the original Spanish and Portuguese trading out of Lisbon and Seville being joined by the French trading out of La Rochelle, the English trading out of London and, most importantly, by Dutch merchants trading out of Amsterdam and a score of other cities. By the mid-seventeenth century it is the Dutch who dominate the overall pattern. Their success was built initially in Europe: first, the Low Countries' longer-term control of the Baltic trade, and second, their successful strategy of blocking off Antwerp from the sea to capture sea/river routes to the Rhineland heart of Western Europe. Beyond Europe the pattern consists of a North Atlantic triangular trading core with further linkages east and west, where Spain and Portugal retained some influence. At the core of the system lay Amsterdam, the 'world's entrepôt', whose commercial calendar was organized around the return of the four great fleets from the Baltic, the Levant, the West Indies and the East Indies (Israel 1989: 257–8). As well as the commodities, information arrived with the fleets and from other parts of Europe to make Amsterdam the first 'world clearing house' for commercial information (Smith 1984). Amsterdam was the financial centre for all this activity and, as noted in Chapter 1, it has been dubbed the hegemonic city of this great city trading era (Lee and Pelizzon 1991).

With respect to the later metageographies, the meaning of the first modern case is particularly interesting. For these first modern Europeans, the rest of the world was viewed as a cornucopia, a land of plenty from which to win great wealth. Hence there was no assumption that these other regions of the world were inferior. This is in keeping with the reality: throughout mercantile modernity, Europe was not the leading world region, as Frank (1998) has so clearly demonstrated. But it was not just the traditional European awe of the 'fabulous East' that is to be found in contemporary minds; European reactions to African cities similarly show none of the later superiority complex: Benin, for instance, was compared favourably to Amsterdam (Oliver and Fage 1988: 89–90). The first modern European metageography was only topological precisely because, beyond the Atlantic, it consisted of merely feeder paths into established Asian city trading networks.

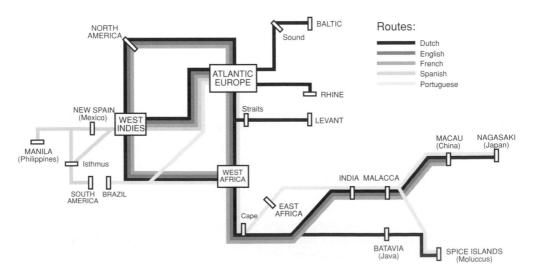

Figure 8.1 The topological metageography of mercantile modernity (from Taylor 2003)

The centripetal metageography of industrial modernity

The Industrial Revolution in North-West Europe changed the balance of power between world regions, thereby generating a metageographical transition. New production, armaments and wealth meant that militarily and economically, Europeans could take over and restructure the world to suit their own ends. In effect, the Atlantic triangular pattern from the topological metageography was writ large to produce worldwide European control. The victory of the British over imperial China in the First Opium War (1841) is the symbolic event that confirms the overturning of the traditional world hierarchy of East over West. The world is no longer a wondrous cornucopia; it now becomes a practical place to be redesigned for new industrial needs.

The resulting centripetal metageography is shown in Figure 8.2. The pattern is extremely simple: there are just two regions, an industrial core, plus the rest of the world supplying the needs of its great industrial cities and towns. The initial development of industrialization in Europe was based upon local access to coal and iron ores, but, beyond these basics, raw materials further afield were soon needed. This required the creation of new production zones, which came in three forms: first, minerals for the engineers (beyond traditional gold and silver to copper, diamonds, nickel, tin, zinc and later petroleum for fuel and plastics); second, industrial crops for the factories (beyond tobacco to cotton, jute, oil seed, rubber, wool); and third, food crops (beyond sugar to grains, meat, fruits, drinks). Of course, these developments took little heed of existing settlement patterns, so that large-scale migrant labour became the norm. Typically, each individual 'island of production' existed in a 'sea of cheap labour': it was a case of labour in from the outer periphery and commodities out to the industrial core. In this way, different parts of the world became specialized in single commodities – so much so that countries came to be commonly associated with particular commodities. Well-known examples are Argentina and beef, Australia and wool, Brazil and coffee, Bengal and jute, Ceylon and tea, Malaya and rubber, New Zealand and lamb, Northern Rhodesia (Zambia) and copper, South Africa and diamonds, Sudan and cotton, and later, the classic case of the Middle East and oil. In the process a core of dynamic cities and a periphery of static cities, as described in Chapter 2 following Jacobs (1984), was created. And at the financial centre of this reordering of world stood London, this second hegemonic city (Lee and Pelizzon 1991) finally taking over from Amsterdam during the Revolutionary and Napoleonic Wars.

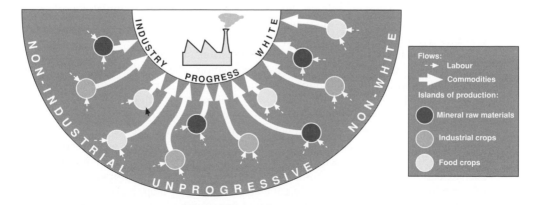

Figure 8.2 The centripetal metageography of industrial modernity (from Taylor 2003)

The meaning of this metageography is straightforward: a simple human hierarchy has been constructed defining Europeans as superior to peoples from other parts of the world. Combined with a temporal metahistory of progress, as typified by the Whig theory of history (Taylor 1996: 133–6), legitimation was provided for pro-white, scientific racism, often integrated with a pro-temperate climate, scientific environmentalism, and a virulent political imperialism resulting in a relatively few whites ruling relatively large numbers of non-whites. In becoming custodians for the unfortunate non-progressives, imperialists did attempt to make the colonies 'pay for themselves', of course, by ensuring that each colony included a specialized island of production. One outcome was the celebrated world map of the British Empire showing pink areas everywhere (Cook 1984). Appearing on the walls of classrooms in all continents, this was the best-known expression of the centripetal metageography.

The mosaic metageography of consumer modernity

Today the world map found on classroom walls is more universal: it shows the boundaries of all states across the world. With independent states replacing empires, the mosaic map is truly a post-colonial artefact. The metageographical transition from which it arose as the latest metageography is decolonization. The first half of the twentieth century saw the gradual demise of Europe as the 'natural' locus of world leadership. Even at the height of the new imperialism, the emergence of non-European world powers at the turn of the nineteenth and twentieth centuries in the Pacific – the United States and Japan – signalled a changing world hierarchy (Bartlett 1984). Japan, in particular, as the first 'non-white/non-European' state to make an independent impact on modern world politics was symbolically important. Although legitimation of European rule was increasingly challenged, political change was quite slow, as reflected in the outcome of the two world wars: after the First World War the area of European imperium was actually increased, whereas the end of the Second World War marked the critical beginning of its final demise.

The world political map that was formed after 1945 has been identified as fundamentally different from all previous international systems (Hinsley 1982). Under the auspices of the UN, the only legitimate wars became wars of defence. In other words, boundaries became sacrosanct. Thus decolonization resulted in new states within colonial boundaries. This boundary maintenance obsession has created a remarkable stability to the world political map: even the later break-up of communist states in the 1990s was kept strictly to existing boundary lines by always making new external sovereign boundaries out of old internal provincial boundaries. The result is a world political map that appears almost 'natural': political boundary lines (red) share a similar visual status to coastlines (black), rivers (blue) and mountain ranges (brown) on maps of the world. It is this status that creates the metageography.

Metageographies are not simple political creations: delegitimizing war by the UN cannot of itself make a new geographical imagination. The mosaic metageography is a product of nationalism. The political boundaries do not just delimit the territories of states, they define the homelands of nations. Combining the political institution of state with the cultural attributes of nation is a nineteenth-century European political movement that created nationalism as a global movement in the twentieth century. The basic nationalist claim is that nations are the natural divisions of humanity expressed through statehood (Smith 1982). Thus nations typically 'find' primeval justification for their presence as nation-states, and it is this constructed 'timelessness' that promotes the illusion of naturalness. Hence the world political map is no ordinary map; it places every viewer, through her or his nationality, within a spatial framework of a segmented humanity. No room here for thinking about cosmopolitan cities and their transnational connections.

The United States is, of course, the first post-colonial state. Pre-nationalist in its origins, it has developed a civic nationalism rather than an ethnic nationalism, with its territory deemed to be a sanctuary for 'timeless values' (freedom and democracy) rather than for a timeless people. Hence despite a minor foray into imperialism at the turn of the nineteenth and twentieth centuries, the United States has been able to portray itself as an anti-imperial champion. During the First World War New York succeeded London as the world's financial centre, as the third hegemonic city (Lee amd Pelizzon 1991). After 1945 the United States was able to harness the implicit democracy within nationalism to create a post-colonial 'free world' open for American business. The subsequent 'American invasion' was by 'multinational' corporations from US cities that viewed the world political map as an array of 'national markets' for consumption of their goods and also as sites for production behind tariff barriers. Of course, it was this great foray into world markets that ultimately created the need for the development of world cities in and beyond the United States (see Chapter 1).

The meaning of this mosaic metageography is subtler than that of its forebears. Of course, political independence did not eliminate the old centripetal pattern, which lived on in concepts such as 'underdevelopment', 'Third World' or simply 'South'. But the mosaic metageography incorporates an inherent sense of equivalence between states that can be expressed as an international egalitarian discourse summed up in the slogan 'development for all' (Taylor 1996: 136–40). In the former centripetal metageography it is only the centre that experiences progress as inherent in superior Europeans; all other civilizations are deemed to be 'stagnant'. Under US hegemony the concept of progress is replaced by development, which is a matter of states, not civilizations. Hence, the promise of Americanization, the American dream as world dream, is promoted as a possibility for any country as long as it employs the correct development policies, as famously modelled by Rostow's (1960) stages of economic growth. Originally this involved 'iron and steel works all round' as 'underdeveloped countries' attempted to replicate nineteenth-century European industrial development. But Americanization gradually converted new political citizenship into an economic citizenship or 'consumptionship' in the famous 'post-war boom'. This was, of course, in line with the needs of US multinationals working in a mosaic of national markets. It is this mosaic that remains important in the global strategies of advertising firms, as shown in Chapter 6.

Embedded statism: living and studying through a metageography

Living with a metageography involves being largely unconscious of its influence on thinking and activities. Its spatial premises enter the realm of 'common sense' where interrogation is deemed wholly unnecessary. The mosaic metageography has fixed an embedded statism into contemporary knowledge of the social world. This is as true of the thinking of social scientists, who are supposed to be critically aware of their environs, as it is of society in general. Certainly there have been important recent statements warning of the perils of unexamined territorial thinking: Ruggie (1993) in international relations and Agnew (1993) in political geography are the most influential examples. However, it is my thesis that these warnings, while being valuable correctives, have not fully understood the intellectual and social depth of the statism that they attack. Quite simply, statism is embedded in society through the mosaic metageography.

Embedded statism has been the unexamined spatial discourse for much of the twentieth century. My favourite example of the overpowering influence of this metageography comes in nature books, where animals and plants are 'nationalized' as in 'British Trees' or 'British Insects', as if the amalgam of 'indigenous species' is somehow uniquely British. There are even books on 'British Birds', even though many are migratory and spend much of the year either far to the north or far to the south of Britain. No matter, they are nationalized in this popular medium, where natural spaces of flows are treated as secondary to human spaces of places. In rather more subtle ways, cities are also nationalized, typically represented as points in a territory rather than hubs within networks.

Nationalization of cities

There are two stages in the incorporation of cities into states in the modern world-system. The first stage is the political centralization of the sixteenth and seventeenth centuries that culminated in the absolutist state of the eighteenth century. In this process, alternative territorial authorities, including vestiges of medieval Europe such as smaller kingdoms and principalities, peripheral magnates and religious sovereignties, as well as cities, were brought more directly under central state control. This spatial reordering created the Westphalia system of sovereign states that is the basis of modern politics. Clearly, such a new politics privileging a space of places at the centre of the modern world-economy was a challenge to the transcontinental space of flows constituting the mercantile metageography. I discuss the effects of this particular state centralization in Chapter 9; all I need note here is that states did not become central to modern society's collective world-image until after they had become nationalized – that is, transmuted into nation-states. This produced the second, and more fundamental, stage of incorporating cities into states and is my subject matter here.

Nationalization of the state increased its power immensely, but, I will argue, nationalization of the cities, in general, undermined the vitality of cities. Why this difference? Basically, nationalization is a territorial reorganization, an assertion of spaces of places over spaces of flows. Thus territorial institutions such as states are privileged; network institutions like cities are constrained. There are exceptions to this rule, notably capital cities, which, as 'national capitals', can prosper immensely from being the co-ordinator and focus of their state's new centralization of power. Hence whereas industrialization created a growth impetus to a range of new cities in the core of the world-economy in the nineteenth century – northern Britain, the Rhinelands, the north-eastern United States – its global effect of changing the military balance of power between core and periphery and subsequent imperialism refocused growth to the capital cities as 'imperial capitals' – London, Paris, Berlin – by the end of the century. The classic case of this process is the so-called 'north–south divide' in Britain.

The Industrial Revolution in Britain occurred away from the political machinations of the mercantile state in London. In fact, many of the leading cities of this economic revolution, notably Manchester, were not even included as boroughs represented in the parliament in London. Nevertheless, in just a few decades either side of 1800 northern British cities became the initial hubs in an industrial network that was to span the globe. This new space of flows involving raw materials flowing in and manufactured products flowing out was not initially a formal imperial system: the Americas beyond Caribbean colonies and Asia beyond the Indian quasi-colony were critically integral. Within Britain the manufacturing hubs in the network were innovative city-regions each with their own specialist niches, such as Manchester/south-east Lancashire and cotton textiles, Leeds/Bradford/west Yorkshire and woollen textiles, Birmingham/Black Country (west Midlands) and metallurgy, Sheffield/south Yorkshire and steel cutlery,

Nottingham/east Midlands and lace and hosiery, Belfast/north-east Ireland and linens, Middlesbrough/Teesside and chemicals, while the great mercantilist ports of Liverpool, Glasgow and Newcastle became centres of industrial zones – Merseyside, Clydeside and Tyneside respectively – focusing on shipbuilding and some processing of tropical goods. Among these great Victorian cities Manchester stands out as the great industrial metropolis, the 'shock city' of its times, attracting visitors from within and outside Britain to marvel at, and learn from, the new world that was being created (Taylor 1996: 124–6). If any city deserves the appellation of world city because of its global impact, nineteenth-century Manchester does.

Within Britain, Manchester's increasing world eminence was reflected in a new political importance. The 'Manchester liberals' were the apostles of free trade and their politics prevailed in the mid-nineteenth century. They did not merely influence Britain but set in place a 'free trade era' that encompassed both European and American states (Taylor 1996: 98–102). Interestingly enough, when the worldwide political tide turned against free trade, creating opposition to it in Britain, it was another industrial city, Birmingham, that provided the political leadership for rejecting free trade in favour of 'tariff reform' and 'imperial preferences'. But this was not a British city's politicians setting a new global agenda. On this occasion it is a reactive politics to others' protectionist policies; their goal is to save British manufacturing from the competition of new foreign industrial cities. The policy reference to the British Empire is crucial here. The late nineteenth-century 'age of imperialism', including a multiplication of British colonies, created a more political orientation to the centripetal metageography. The result was an enormous privileging of capital cities. In the case of Britain this led to a huge downgrading of northern cities from major world status to minor 'provincial status' (provincial to London) in a few decades either side of 1900.

In the 1890s south-east England, the region around London, acquired a new name: the Home Counties. This was the imperial centre, the place from which men left to win and run the empire and afterwards to where they retired. Thus it was their 'true home' throughout a career that might encompass many different parts of the world. London was the epicentre; aspiring to being more than the national capital, it was a grand imperial capital housing an imperial parliament. This new imperial Britain eschewed its industrial heritage as nasty, dirty and brutish, not in keeping with a great imperial people. In this way British cities, apart from the capital, were written out of the new national ideology, which basked in rural images of the Home Counties (Taylor 1991). Indeed, an exclusive Englishness became the dominant motif of the state, through which the Home Counties became transmuted into the effective 'homeland' of this imperial people. Thus the twentieth century was marked by a 'north–south divide' across the country that separated a dystopian industrial urban north from a utopian rural south. In the first half of the century the old industrial cities continued in the role of milch cows for the imperial cause, especially in two world wars fought using their industrial productions, but in the second half of the century, in particular, they were reduced to state-dependence status in the wake of a virulent de-industrialization. Certainly, British imperial nationalism has not been good for British industrial cities in the era of the mosaic metageography.

Jacobs' (1984) description of how a single leading city gains from national macro-economic policy, outlined in Chapter 2, is a separate economic process to the political process described above. In combination, they create an extreme primate city pattern within the state. Of course, capital city privileging, while particularly notable in Britain, has been a general political process, albeit taking different forms in different countries. The end result is that apart from in a few large countries, the primate city pattern was nearly universal in twentieth-century urban development.

In complete contrast, some states have become city creators, often to counter the primate growth of their leading city. They apply spatial policies that explicitly favour the territorial needs of the state over the network needs of existing cities by channelling investment into a new capital city. It is as if states are jealous of their leading cities: by pre-dating their state (London is older than England, Stockholm is older than Sweden, New York is older than the United States, Sydney is older than Australia, Toronto is older than Canada, etc.), most leading cities of the world illustrate their incipient independence. In fact, there are two pragmatic political reasons that have governed capital city movement, including new city creation. Initially, most such spatial interventions in the national urban network were national compromises so as not to alienate one or other section of the nation. Washington, DC, Ottawa and Canberra are examples. More recently, capital city movement has been part of a national development strategy. After state revolutions, either decolonization or profound regime change, the old capital is deemed to represent the old ways. These old ways are typically economic sinks, the locales through which external powers extracted wealth from the colony or old regime. Because of their function these were usually at the edge of the national territory. The nationalist reaction to this has been to relocate the political capital more towards the centre of the national territory to produce a 'purer' national city. Early examples are the Bolsheviks' returning the Russian capital from St Petersburg to Moscow, and Atatürk moving the Turkish capital from Istanbul to Ankara. More recently, Brazil and Nigeria have moved their capitals inland from the coastal cities of Rio de Janeiro and Lagos respectively.

These are all examples remarkable for the fact that the states think they can create cities. Indeed, they can't; creating cities that are designed to be national and not cosmopolitan is a recipe for urban disaster. And so it has worked out: of all these cities, only Moscow is currently its state's leading city, but this is an unusual case in that it was the original capital of the state. Of the rest, only Washington rates at all as a world city. States as territorial entities are not good creators of network entities, as evidenced by their city development disasters. The many world cities that are also capital cities have harnessed state resources alongside their network resources: they are national *and* cosmopolitan.

Nationalization of the study of cities

Although there is a long tradition of pondering the relations between people, peoples and their environments, social science is a relatively recent invention. As an idea it is a product of the nineteenth century, but it came to full fruition as acceptable disciplinary knowledge in universities only in the twentieth century. It was constructed on the frontier between the much more established knowledges of natural science and the humanities. Based on the simple premise that it was possible to use the methods of the former to study the subject matter of the latter, social science was developed initially under the pervasive influence of social progress and reform. Originally there was a flowering of new separate social knowledges in orthodox political economy, positivist 'social science' and liberal political philosophy. This early social science directly reflected the centripetal metageography of its time. Its theories and models were only designed as applicable to 'industrial society'; the rest of the world, being non-progressive, was intellectually uninteresting and left largely to the curiosity of anthropologists, geographers, Orientalists and their ilk.

Subsequently, political economy has transmuted into economics, positivist 'social science' into sociology, and political philosophy into political science. Although building upon European antecedents, this 'trinity' of contemporary social sciences was finally created and promoted through American academe in the mid-twentieth century. I use the word 'trinity' here not just in the quantity sense but also to indicate a collective mutuality. All three disciplines have maintained some of their forebears' broader knowledge claims, but essentially they have become more specialized, concentrating on 'economy', 'civil society' and 'politics' respectively. And this is where the mutuality comes in: between them they claim to study all social relations, leaving no intellectual space for other disciplines.

The social science trinity also captured all geographical space for themselves. As the centripetal metageography declined, so the old 'non-industrial' disciplines, such as anthropology, began to lose influence in the realms of academe. With the coming of the new mosaic metageography, the trinity was expected to provide knowledge of all countries equally; it obliged by simply transferring knowledge derived from 'developed' countries to 'underdeveloped' countries. The result was the creation of a variety of modernization and development theories for, but not of, 'Third World' states. And, of course, this new social unitarianism extended to time. Taking a particularly narrow view of the science it was importing into the study of human activities, twentieth-century social science at its worst offered timeless and spaceless knowledge based upon universal social theory.

But there was a hidden premise in this universal theory. Embedded statism contains the remarkable geographical assumption that all the important human social activities share exactly the same spaces. This spatial congruence can be stated simply: the 'society' which sociologists study, the 'economy' which economists study, and the 'polity' which political scientists study all share a common geographical boundary, that of the state. However abstract the social theory, it is national societies that are described; however quantitative the economic models, it is national economies that are depicted; and however behavioural the political science, it is national governance at issue. From a geographical perspective the trinity provides 'one-scale' social knowledge as directed by the mosaic metageography.

Since the real world did not actually consist of homogeneous and autonomous mosaic cells, the study of social practices above and below the privileged scale could not be wholly eliminated. However, the state remains as a sort of intellectual pivot in these other studies at 'international' and 'subnational' scales. Inter-state relations have been studied largely through a realist International Relations that eschews any social relations in its war/preparation-for-war prescriptions. Below the state, a few economists have focused on regions, sociologists have studied communities and political scientists have analysed local governments. Cities feature in all these trinity knowledges but they are not usually the focus. The latter obtains in an urban studies that is explicitly interdisciplinary, or even transdisciplinary, in nature, a potential threat to the trinity to the degree that it reveals its statist limitations. But the threat has rarely materialized: as I showed in Chapter 1, urban studies may be transdisciplinary but it has not been transnational in content.

Thus although studies that identified 'national urban systems' were important (and since, overall in social science, studies of inter-city relations are quite rare, they should be celebrated) they resulted in an urban studies that did not escape the mosaic metageography. This shows the power of the knowledge straitjacket: even studies of relations between cities stopped at the national boundary (Bagnasco and Le Gales 2000: 29). Theoretical arguments such as the primate city model and the rank-size rule described in Chapter 1 conceptualized processes at a national scale only. The national urban hierarchy, at least when it fitted the rank-size rule, was the outcome of a system equilibrium, and the urban system was just as national as the economist's economy, the sociologist's society or the political scientist's polity. The study of cities had been nationalized.

There are, of course, many good reasons for developing social knowledge at the scale of the state. The rise of the mosaic metageography relates to the rise of states as the prime actors in social affairs. No longer just remote vehicles for collecting taxes to fight wars, in the nineteenth and twentieth centuries states grew immensely in servicing their citizens, 'from cradle to grave', as the saying goes. This is the nationalization of states previously discussed. For instance, national currencies led to central banks through which national governments could pursue macroeconomic policies of demand and/or supply management in the national economy. Similarly, any national regional planning requires a national city system to operate through. Clearly, any meaningful social knowledge of these times must have the nation-state as a prime subject. But if it is to be a critical social knowledge it must also appreciate that the nation-state is much more than a subject. Not only the social science trinity, but also urban studies has failed to transcend embedded statism; the state is unexamined at the very heart of the social knowledge project: how the knowledge is organized.

The nationalization of urban studies is so shocking because the object of study is a network entity. National urban systems were just as much a myth as national economies because throughout the modern world-system there has existed a world-economy built upon both international and transnational processes. For instance, as Arrighi (1994) has clearly shown, international financial flows transcending states have been a cyclical phenomenon throughout the history of the modern world-system. Interestingly, Arrighi suggests that Castells (1989) misses earlier spaces of flows because he is a victim of embedded statism in his treatment of pre-1970 spaces; Arrighi (1994: 84) neatly calls the process 'the bias of our conceptual equipment in favour of space-of-places that defines the process of state formation'. Cities have continued to import and export within and beyond their territorial state boundaries throughout the period covered by the mosaic metageography; it is just that the political ordering expressed in the latter has thrown a perceptual veil over the transnational economic dynamic that sustains the politics.

If we return to the primate city model and the rank-size rule that dominated theoretical thinking on inter-city relations in the 1950s and 1960s, it can now be appreciated how such ludicrous modelling – London at the top of a primate city model and New York at the top of a rank-size city graph, as if these cities' critical *raison d'être* were not to be found in relations outside as well as inside their respective states – could come to be widely accepted. As is argued in Chapter 1, the world never consisted of 100 or more separate 'urban systems'. All primate cities and the cities they dominated, all cities wherever located in rank-size graphs, were never wholly national in their connections. Many, as ports and latterly as airport hubs, were primarily engaged in extra-territorial business. With such permeable boundaries, the possibility of any credible systems analysis at the scale of the state has never been feasible. Within the inter-city networks of the time, boundaries were not unimportant to economic transactions, but they were not the defining characteristic. Rather than assuming the latter, boundaries required studying in terms of what effect they did have on transactions. There were rare examples of the latter, notably by Mackay (1958), in which boundaries were calibrated in terms of their 'distance effect': putting a political boundary between two cities is placing a constraint on their inter-city trade that can be thought of as increasing the distance between them (see Figure 8.3). Thus, in effect, the world mosaic of state boundaries made the world much 'larger', in the sense of increasing transaction costs within the world-economy. It is just such transaction costs that, in the sectors I have focused upon, have been greatly eliminated as direct cost, but with indirect costs (culture) remaining, as particularly shown in advertising firms' global locational strategies discussed in Chapter 6.

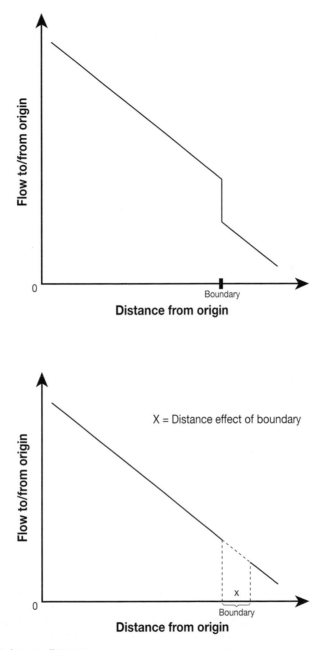

Figure 8.3 Boundaries as distance

Globalization as a metageographical transition

A metageographical transition is an alteration in the collective geographical imagination. It marks the disintegration of an existing metageography and provides the conditions to create a new metageography. Such a transition is not, of course, instantaneous,
but there should be a clear-cut change in the way the world is viewed and interpreted.

Thus in considering the present as a metageographical moment I should be searching for evidence of an erosion of embedded statism, of a serious decline in acceptance of the state as the 'natural' locus of power across the gamut of human social activities. At even the most cursory level of consideration, globalization seems to fit the bill as a metageographical transition.

At its heart, most of the debate about globalization has been concerned with the future of the state. For instance, David Held and his colleagues (1999: 2–10) align positions on globalization along a spectrum with the 'hyperglobalists' at one end and the 'sceptics' at the other. The former define a 'global age' and the end of the state, the latter enhanced 'internationalization' with states remaining as key actors within regional blocs. In between are the 'transformationists' who accept the unprecedented levels of global interconnectedness but view this as transforming state power, not eliminating it. Part of the problem with this argument is that the terms of the debate were originally set by the hyperglobalists, and this has produced a global versus state agenda. The transformationists try to free themselves from this simple opposition but in doing so they deal more with the transformation of the state than with global transformation itself (Taylor 2000b). In contrast, my starting point is the question of the geographical scale at which human social activities take place.

There is one thing all positions in the debate are agreed upon: there have been changes in the pattern of geographical scales at which human social activities are organized. To be sure, the degree of change is in dispute, but even the sceptics challenge the simple one-scale pattern of human organization ensconced in the mosaic metageography. I focus on the transnational processes within globalization. Boundaries are about truncating flows, and this is precisely where states are being challenged as new technologies have made policing borders more difficult than ever before. The World Wide Web is, in old terminology, a 'smuggler's charter'. This is not to argue the hyperglobalist position of a 'borderless world', but it does mean that transnational processes can undermine the mosaic metageography much more completely than at any time during its existence.

As I described in Chapter 1, for Castells (1996) the world city network represents a new critical layer in his network society. Is this new society critically urban? I am going to answer this question positively, rehearsing arguments made earlier. Economic restructuring associated with globalization has enhanced the role of cities to an extraordinary degree. The informational age has favoured the growth of economic sectors where knowledge (i.e. how to use information) is a key component of production. In advanced service sectors, of course, knowledge is the product. Hence the traditional role of cities as 'service centres' has meant that the major cities that provide corporate services have prospered outstandingly with globalization. These advanced producer services have 'gone global'. Initially, this locational expansion was necessary to serve their multinational corporation customers, but subsequently service firms have developed their own global strategies. Making full use of the enabling communication technologies, advanced producer service firms themselves created a cutting-edge global industry. Hence service firms, traditionally associated with a client base in a single city, have become global corporations. But to stay at the cutting edge they have to be located where large quantities of the most up-to-date information intersect with clusters of advanced 'professional/creative' expertise to make new customized knowledge products (Sassen 1991). This means that to be a successful player in the field, a global service firm has to have an array of offices in world cities across the globe.

Thus the many maps I have produced in the previous chapters represent a different world from that normally derived from social science research. Eschewing the mosaic metageography, I have produced an overall picture of what the geography of Castells' (1996) network society might look like. Can or will this world city network constitute a new metageography in keeping with contemporary globalization? The answer to 'can' is obviously yes: with the relative disintegration of the mosaic metageography as described earlier, the condition exists for the making of an alternative collective geographical imagination. But this is not a simple and quick switch. In answer to 'will', my confidence must be highly circumscribed. As things stand at the moment, cities appear to be the most likely candidates to threaten the primacy of states as the building blocks of a spatial framework of the world. But there is, as yet, little or no evidence of a transfer of primary allegiance or identity from state to city. The jury will be out for quite a long time before it will be possible to assess whether globalization and the concomitant rise of world cities is or is not a metageographical transition. The next chapter explores this possibility alongside other futures.

9 From present to future: reasserting cities?

> So, the geography of the new history will not be made, after all, of the separation between places and flows, but out of the interface between places and flows and between cultures and social interests, both in the space of flows and in the space of places.
>
> (Castells 1999: 302)

Looking into the future is supposition without the possibility of any direct empirical back-up. Nevertheless, it can be argued that the ultimate *raison d'être* of any social science study is to provide knowledge of the past and present to aid in understanding how to create a better future. To resolve this paradox, social scientists extrapolate using a range of mixes of empirical and theoretical knowledges to provide informed speculation or 'best guesstimates', both utopias and dystopias. In Chapter 8 I identify the present as a putative metageographical transition called globalization in which cities are playing a key role. Here I accept the veracity of this interpretation and look at its implications for the future. However, I am not a futurologist; I do not offer extrapolations in the form of my own predictions of a possible city-centred future. Rather, this chapter brings together some ideas for trying to understand a future in which cities are becoming increasingly important.

I begin this chapter with two future scenarios from the literature that locate cities at their centre. One is a dystopia, the other a utopia; between them they define limiting arguments in the range of possibilities. They provide an initial flavour of city-centred futures; I use them to set the scene for my own discussion. For the latter, I begin in the second section with a review of the new evidence from the global urban analyses on cities and globalization at the beginning of the twenty-first century. This evidence is selectively rehearsed in conjunction with theoretical ideas from Jacobsean economics and world-systems analysis to provide what I understand to be my 'basic knowledge' for understanding future possibilities. In the third section I take a glance back to the 'golden age' of the Dutch in the seventeenth century. This may appear odd in an argument looking to the future, but without any direct empirical props I find that I need to find a 'pattern state', a polity whose social matrix can inform city-centred thinking. The final section attempts to go beyond both place-based knowledges of political economy in general, and the idea of 'city as crucible' in particular, to explore network-based understandings for future social possibilities. This is to excavate deterritorializations of such key social practices as law and democracy. In the process, a serious revision of world-systems analysis's geohistory emerges. This extention of supposition through into the initial theoretical global framework is where this chapter and the book conclude.

Two city-centred future scenarios

To begin my peering into the future I will briefly review two scenarios that address some of the themes of this book. There is, of course, a long tradition of books (and other media) that deal comprehensively with future cities (e.g. Blowers *et al.* 1974; Brotchie *et al.* 1995), but few writings consider future *inter-city* relations. The two scenarios I have chosen provide extreme possibilities for future cities in globalization, creating both a global dystopia of cities and a city utopia of freedom. These provide me with limits for my thinking, harbour lights for me to steer between in trying to understand future possibilities for the world city network.

Petrella's corporate dystopia

Riccardo Petrella (1995), described as 'the official futurist of the European Union' (p. 21), portrays a chilling city-centred future in which today's G-7 leading states are replaced by a CR-30, the top thirty city regions. These cities will dominate the world through an alliance of the 'global merchant class' and metropolitan governments, 'whose chief function is supporting the competitiveness of the global firms to which they are host' (ibid.). This scenario is based upon a hierarchical arrangement in which competitiveness becomes a 'battle between city-regions' rather than between nation-states (ibid.: 22).

The world politics of this dystopia is an inclusive/exclusive set of processes that makes the traditional core–periphery model look like a good deal for the poor. The CR-30 will be constituted by rich city-regions with 'manageable' populations of between 8 and 12 million. These world cities form 'one giant web', a 'wealthy archipelago of city regions' separated from an encompassing impoverished world (Petrella 1995: 21). This excluded *lumpenplanet*, to use Petrella's terminology, consists of poor city-regions of largely displaced peasants with huge populations of over 15 million. Thus this is a political world in which world cities face mega-cities. The latter will be the primary source for propagating a series of global criminal activities (spaces of smuggling flows ranging from drugs to people), to which the former will respond by erecting barriers to entry. Thus 'gated city-regions' will localize the 'good life' while the rest of the world sinks into an abyss of poverty and violence.

What is the world geography of these processes? Interestingly, this extreme exclusiveness does not leave out contemporary non-core world cities such as Johannesburg, Shanghai, Jakarta, Kuala Lumpur and São Paulo. The old core–periphery remains as little more than a shadow reflected in there being more CR-30 cities in the 'North' than in the 'South'. Thus this scenario is not a new way of shoring up the established core zones of the world-economy; it represents a new future geography in which the core is formed by 'disassociated islands' rather than by contiguous zones. Furthermore, the dystopia cuts out the semi-periphery: in Petrella's scenario there is no hope for change; the process of rising up through the world-economy is simply locked out of this stark new geography.

Although Petrella (1995) posits an alternative positive view of a city-based future based upon the rise of a new global civil society, it is his pessimistic view that has attracted attention. This is because, although it is a worst-case scenario, it actually chimes with many contemporary trends. For Petrella it is the 'most likely scenario in the immediate future' (p. 21), but he thinks it to be unsustainable in the longer term. This may well be; here I use Petrella's ideas as one extreme possibility for future cities in globalization.

Jacobs' 'utopian fantasy' of secession

Whereas in Petrella's dystopia, cities, the CR-30, are the problem, for Jane Jacobs (1984) it is the actions of states that threaten future well-being. As detailed in Chapter 2, states constitute an amalgam of economic spaces, both city-regions and other less fortunate economic regions, a spatial arrangement that is by no means simply benign. For Jacobs there is always pressure within states to 'kill the goose that lays its golden eggs' – that is, cities. Jacobs confronts this state–city predicament and suggests a solution: city secession.

Jacobs (1984: 214–15) suggests that it is economically sensible to replace the traditional state concern for territorial integrity with a propensity for a 'multiplication of sovereignties by division' (p. 215). It should be the normal response to economic development within a state in recognition of increased complexity of city relations. With different cities in different stages of economic cycles, it makes economic sense to separate. This is not a recipe for civil war because separation will benefit both the original state, as it sheds its declining city regions, and the 'young sovereignty' that gains some control of its economic destiny – for instance, through having its own currency. Thus all divisions would be friendly and consensual in a gradual transformation of territorial states into a new geography: from a mosaic economic world to city-states in a nodal one. Although these nodes would, at any one point in time, be at different economic cycle stages, this geographical differentiation would not take a core–periphery structure. The division of a world city network into dynamic cities and static cities would become a contingent matter, perhaps a fluid geography with particular mixes of core-like city-states and semi-peripheral-like city-states at any given time.

This is, of course, in Jacobs' words, a 'utopian fantasy' (ibid.: 215), because states – nation-states – are not only economic spaces. Because nations are forged by military and political processes, division is anathema to national governments, whose first duty is to preserve the unity of the nation. To do otherwise is a betrayal of all the blood sacrifices of past generations through which the nation has been created (ibid.: 216). Thus peaceful partitions of territory are rare: Jacobs mentions the separation of Norway from Sweden in 1905 (ibid.: 218), and the more recent case of the division of Czechoslovakia into the Czech Republic and Slovakia can be mentioned. Perhaps the most significant example is Singapore's separation from Malaysia immediately after decolonization to produce a city-state. Also, although the circumstances of Hong Kong have changed, as first a British colony and latterly part of China, this city-region maintained a high degree of economic autonomy in both its political subordinations. The economic successes of these two single-city 'sovereignties' have been nothing short of spectacular and add weight to the ideas that lay behind Jacobs' 'utopian proposal' (ibid.). But these successes involve just two of the world cities that constitute the world city network. The rest remain hitched to territorial states, mere parts of aggregated spaces of economic policy making, albeit with some cities dominating their 'national economy' as 'favoured primates', as described in Chapter 2. Of course, secession is not a panacea for regeneration of both old and new polities; it just provides a geoeconomic framework to make it more likely. Dynamic cities still have to be created and sustained.

For Giddens (1998: 129), 'a world of a thousand city-states' is a recipe for chaos. This may well be, but for my purposes here I will interpret Jacobs' utopia as the other extreme possibility – utopian freedom against dystopian dominance – for future cities in globalization.

Cities in globalization: some basic knowledge

The basic finding of the global urban analyses has been to show that there is indeed a world city network and that this network has a particular geography that will inevitably impinge on future social change. However, in this section I do not attempt to provide a summary of the many results and findings reported in earlier chapters. Rather, I am very selective and draw upon empirical analyses to explore how new knowledge of cities in globalization informs the theoretical underpinnings of this research. In other words, I am returning to themes broached in Chapter 2 as 'back to basics' to reassess their meanings in the light of the global urban analyses. This involves engaging in three key contemporary questions that have profound resonance for the future of cities and of world society. Following Petrella, I ask: how do cities in globalization relate to world-wide economic polarization? Following Jacobs, I ask: in what sense are cities in globalization becoming more autonomous? Finally, I highlight the most surprising empirical cluster of findings in the global urban analysis, as far as I am aware not presaged anywhere in the urban studies literature: that is, the American exceptionalism within world city network formation. The question is: how will cities in globalization develop in a world-economy dominated by a single, territorial 'superpower'? All three questions address the issue of how a global space of flows interacts with an international space of places.

Core–periphery in the world city network

Contemporary world geography is dominated by two powerful economic tendencies. On the one hand, the world is becoming more and more economically integrated. International financial markets are the virtual epitome of this integration and the development of the world city network is its prime geographical expression. On the other hand, the world is becoming more and more economically polarized. The widening income gap between the richest and poorest percentiles of the global population is the key statistic; the economic decline of a whole continent, Africa, is its prime geographical expression. Furthermore, the two tendencies are broadly synchronized. The literatures on both rising integration and rising polarization generally chart their origins from the 1970s, culminating in a *fin de siècle* world economy that is arguably more integrated and more polarized than ever before. This is the globalization conundrum; it is what makes Petrella's dystopia appear so chillingly possible.

There are two basic state-centric interpretations of this conundrum: it is either an unfortunate chance effect creating an odd paradox or it is a fundamental causal nexus creating a negative outcome. For orthodox economics, integration – 'free' markets – is good for all and will inevitably lead to poorer countries 'catching up' as they 'develop' like richer countries. The fact that the opposite has been happening for a quarter of a century must therefore be a non-causal coincidence due to non-economic factors (the blame is largely attributed to the indulgencies of poorer states). Thus the 'real world' (geographical evidence) is not allowed to interfere with the predictions of economic theory. Of course, Pacific Asia is the only 'Third World' region to show sustained signs of catching up, and countries here have been notorious precisely for *not* freeing up their markets! The alternative position is that it is the economic integration that is actually creating the conditions for generating economic polarization. Freeing up markets damages the prospects of poor countries: an 'open world' is inevitably a polarizing world. Despite the 'real world' (geographical evidence) supporting this position, such heterodox 'Listean' economics ('state economic protectionism' in either nationalist or radical form) is arguably less influential, in both theory and practice, at this *fin de siècle* than ever before in the modern world-system.

How does a more city-centric approach deal with the globalization conundrum? In many ways, the world city network and the core–periphery model represent the two sides of this conundrum: integration operates through the network, polarization accentuates core–periphery. Thus I will answer the question by returning to the matter of how the world city network relates to the core–periphery pattern in the world-economy. The basic geographical issue is that whereas world city network formation is clearly a core-like process involving high-tech production using high-value labour, it is also a process that appears to be happening as a worldwide phenomenon. Thus while the core-located cities of London, New York, Paris and Tokyo are universally regarded as world cities, it is also widely recognized that Mexico City, São Paulo, Istanbul, Johannesburg, Bangkok, Mumbai, Singapore, Shanghai, Taipei, Seoul and Hong Kong also have world city status: the urban global analyses in earlier chapters show them all to be prominent in the world city network. Quite simply, these 'former Third World cities' are integral to the office location policies of important global business service firms in their need to provide worldwide provision to clients. As argued in Chapter 4, such cities have become powerful 'places to be'. How can this network integration be related to the core–periphery dynamic of world-systems analysis? There are three ways in which this question has been answered and I list them here in order of the degree of challenge they pose to conventional interpretations of the core–periphery model.

The simplest interpretation of 'former Third World cities' within the world city network is to invoke the semi-periphery category: these cities are experiencing global core-producing processes while simultaneously incorporating peripheral-producing processes as part of their local regional context. This was how Friedmann (1986) originally framed his 'world city hypothesis' with his table listing world cities divided into core and semi-periphery categories. Thus for Friedmann (1986, 1995), the former Third World cities represent semi-peripheral outcomes with their overt mixture of core (world city) and periphery (e.g. mega-city) processes. This is to interpret the development of a world city network in classic world-systems analytic terms. In the context of contemporary globalization, the rise of world cities is a continuation of the formation of semi-peripheral zones, theorized as the most dynamic part in the world-economy. For this interpretation, world city network formation within globalization does not signal a fundamentally new process; rather, it is a new configuration of an old, long-running unfolding of the modern world-system, the rise of semi-peripheral spaces during major world-economy restructuring. Contemporary restructuring features a new emphasis on spaces of flows but this does not alter the fundamental core–periphery model underlying the world-economy.

The second interpretation is a much more radical use of the core–periphery model. In his study of Santiago as a world city, Jones (1998) interprets the existence of a 'globalised business elite' within the city as indicating a need to 'retheorise the core'. In this argument all world cities are core areas, so globally the core consists, at least in part, of a widely dispersed distribution of places. This is a departure from the world-systems finding that historically core and peripheral processes are geographically concentrated and occur in broad zonal patterns across the world. According to Jones' interpretation, the contemporary core is a series of 'economic islands' of relative prosperity, a 'core-archipelago' rather than core-zone. This is a world-systems interpretation of Petrella's dystopia for world cities but without his mega-cities consequencies. Since, in world-systems analysis, the zonal pattern is a historical but not a necessary element of the core–periphery model, it follows that the archipelago represents a new configuration of polarization and not necessarily a challenge to core and periphery as processes.

The third interpretation is a radical departure from core–periphery thinking. According to Sassen (2000: 151), the 'new geography of strategic places . . . cuts across national borders and the old North–South divide'. This opens up a whole new politics 'along bounded "filieres"' rather than through territories and zones. Again there are shades of Petrella's dystopia here, and in fact Sassen's ideas can be extrapolated to the notion that the world city network has superseded core and periphery. Thus the cities in globalization are integral to a unique contemporary period of change. This would imply Hopkins and Wallerstein's (1996) 'age of transition' rather than simply interpreting globalization as a metageographical transition.

I think there is some truth in all three of these positions and, of course, it is not possible to draw conclusions about their relative merits until globalization processes, and world city network formation in particular, have operated through more than one phase of world-systems development. Thus far, world city network formation has operated only in a single major phase of the world-economy, the Kondratieff downturn that started in the early 1970s. It can be noted, however, that there is no necessary contradiction between network and zone in world-systems geography. Core and periphery are both processes and therefore the zones they dominate (give their name to) should not be viewed as formal homogeneous regions, rich and poor respectively. Rather, they are functional heterogeneous regions, each with comprehensive class structures, so there are always many rich people in the periphery as well as poor people in the core. The difference is in the balance of economic strata between the two zones. From this perspective the fact that world city formation includes cities dispersed across the world is unremarkable: all past economic restructurings have created 'new rich' in both core and periphery; in the contemporary case these creations have been particularly city-centred. But the key finding of the global urban analyses is not just about the global configuration of world cities; the identification of new locales of power created through the network is vital. As is shown in Chapter 4, while command power remains resolutely in core-located cities, the creation of a worldwide network of cities diffuses another sort of power. This network power is found in non-core cities that have become integral and essential to the servicing of global capital. This is a new circumstance, totally different from 'Third World' cities as entrepôts in, first, imperialism and then, Frankian dependency theory. It creates a world city conundrum: the network has been responsible for a dispersion of economic power in a polarizing world of increasing concentration of power. This conundrum may be viewed as a particular form of the globalization conundrum.

To return to the latter, the important element of the argument so far relates to the timings: economic globalization, like the world city network, traces its origins back to the new international division of labour in the 1970s, which means that its processes have operated, and its impacts unfolded, largely through a long world-economy downturn into the 1990s. It is in this context that polarization should be understood. Historically, such downturns have been associated with increases in economic polarization. The over-production/under-demand that creates the need for restructuring favours capital over labour (Wallerstein 1984), and flexible capital (especially financial capital, see Arrighi 1994) in particular. Basically, this is capital that can move easily between sectors and places to attain the highest level of accumulation, thus gaining at the expense of the rest. In the contemporary world-economy, cities are involved directly in this polarization process as service firms have become cutting-edge technology enterprises and business services have become central to the restructuring. In the case of the world city network, the fact that some power has been diffused is totally swamped by these larger economic forces that have steered contemporary globalization.

This indictment of world cities and the meagre distributive capacity of their network must not be the last word. To evaluate a major change in the world-economy over a single economic phase is to invite misguidance. The need is to assess the world city network within globalization in the context of a world-economy growth phase, decades of alternative constellations of economic power where consumption catches up with production. In such circumstances, will the network power of cities be a part of depolarization processes and an expansion of the semi-periphery? These are the type of questions that cannot be answered with the evidence that is available: the contemporary world city network is a creature of globalization as a package of polarizing practices and therefore there is no experience of it in other economic contexts. Furthermore, the theoretical frameworks I am working through (Chapter 2) do not provide the tools for making predictions about the role of a world city network in new economic circumstances. Thus the future I discuss in this chapter will perforce be rather more general. I will assume that the world city network as part of globalization processes is not inherently regressive in nature. The world city network can do more than service global capital and, in the long run, this is what may be of critical importance, as will be explicated later.

Globalization and the 'freedom' of cities

In the traditional studies of inter-city relations described in Chapter 1, the purpose of the national urban hierarchy is to spatially integrate the 'national economy'. However, if the latter is deemed a myth, where does this leave the idea of a national urban system? As noted in Chapter 2, for economic growth to develop through import replacement, it makes no difference whether a city is replacing commodities from foreign or domestic cities (Jacobs 1984: 43). But this does not mean that the idea of a national urban hierarchy is also a myth. There may be no economy for the cities in a country to integrate, but, as has been shown, a primate hierarchy can evolve as a product of the economic policies in a state's economic space. The processes favouring one city over others will produce an outcome that does appear to produce a national hierarchical structure. Thus the national urban hierarchies, ubiquitous to the traditional urban studies reviewed in Chapter 1, can be seen, in part, to be artefacts of the myth of 'national economies' through economic policies embedded in that myth.

This interpretation turns the idea of a national urban hierarchy upside down: instead of cities economically integrating nation-states, it is cities that are being politically integrated into nation-states. This is a vital distinction. Doing the integrating requires an exercise of city power; being integrated implies a loss of city power. This illustrates better than any other example the stark contrast between a state-centric and a city-centric approach to understanding cities. Historically, in the state-centric story, cities are scripted as integrating the national territory through a series of transport innovations; in the alternative reading, cities are being viewed as contained by the state, with their cosmopolitan heritages and potentials being nationalized.

Bringing globalization into the argument can now be read as a 'freeing' of cities from containerization imposed by states. Marrying Jacobsean economics to world-systems analysis entails this 'transnational' view of world cities. London is not simply Britain's world city, any more than Milan is Italy's world city or Toronto is Canada's world city. These cities and other world cities operate through a world city network that is reproducing cities beyond their state's exigencies. Obviously, states are not disappearing from city life, but their threats to the well-being of cities, which so concerned Jacobs (1984), are not what they used to be.

Specifically, what are these state threats? Jacobs (1984: 212) considers nation-states to be at best 'chancy' environments for the future prosperity of cities (and, by extension, of modern civilization), and at worst they 'become lethal weapons'. Her doubts are certainly borne out by history. For instance, cities were early modern victims of the state centralization processes that created our modern state. And to resist was to tempt fate in a big way: in 1585 the sacking of Antwerp by Habsburg forces famously led to the flight of the leading entrepreneurs, marking the end of the city as a leading economic force. Recent history is not that different. In the Cold War the nuclear weapons of both main protagonists were not simply targeted at each other's country, but were aimed specifically at their respective enemy *cities*. In the words of William Bunge (1988: 28), a 'one bomb, one city' policy meant that each country planned not just to decimate the enemy, but to 'decapitate' its nation – that is to say, knock off the top of its urban hierarchy so that the receiver of the missiles could no longer function as an integrated spatial unit. Fortunately, these well-rehearsed military plans never came to fruition, but in the debris of the Cold War, cities again have become victims. Sarajevo was unfortunate to be the capital of Bosnia during the latter's civil war, being regularly bombed by Bosnian Serbs from the hills around the city. It is clear that cosmopolitan Sarajevo would be better off without a political entity called Bosnia. Similarly, the sophisticated city of the Danube basin, Belgrade, had to suffer bombings from US aircraft as the capital of the territorial rump that was Yugoslavia. Significantly, the bombing cut off the key connections of the city, its initial *raison d'être*, by blocking the Danube. And, of course, al-Qaeda's attack on the United States was anything but random, successfully targeting the cities of New York and Washington, DC, as the economic and political hubs of the country. Political enemies can certainly be very lethal as far as cities are concerned.

But Jacobs (1984) is not just thinking of military physical destruction when she voices her concern for leaving cities to the mercies of states. In fact, her main concern is what she terms states' 'transactions of decline' or 'killers of city economies' (p. 183). These come in three forms: massive military expenditures, unproductive subsidies to poor regions, and promotion of unequal trade between advanced and backward economies. These policies, she argues, are as important as the faulty feedback mechanisms, described in Chapter 2, for economically harming cities. The unequal trade policy, for instance, was the outcome of nineteenth-century European imperialism that sapped the competitive edge from British cities in particular. Jacobs (1984) argued that the effects were still to be seen through to the 1980s in the political machinations of the European Community. She interpreted European politics as being 'to a considerable extent the jockeying of two played-out empires, France and Britain, over which one can more successfully milk the cities of Europe to temporize with its own decline' (p. 202). She argues in general that 'cities are the milch cows of economic life' (ibid.: 106). More contemporary examples would be: sapping the vitality of Japanese cities by rural-based political party machines, and the threat of military-economic voraciousness to contemporary US cities.

In what sense are these 'killers of cities' declining under conditions of contemporary globalization? Certainly they are not disappearing, as the previous examples testify, but equally it is possible to discern an emerging freedom of cities. World cities – all world cities – need networks that are global and therefore intrinsically transnational. It is in this context that state custodianship of cities is breaking down. Although states remain important as part of contemporary globalization, there are, nevertheless, important parts of the workings of cities that are being fundamentally liberated from entrapment within territories. Cities can reach their wealth potential only through operating in a space of flows, and imposition of boundaries as barriers to those flows is harmful. With globalization, certain parts of cities' economies are becoming less and less accountable to the encompassing state. The prime example is the financial markets, in which currency trading, for example, has undermined state control of what has been a key indicator of state sovereignty. But there are very many other important economic city transactions that are conducted globally, as the world city literature has illustrated in great detail, not least the business services that have been the focus of my global urban analysis. In all cases, states continue to play a part, as described in Chapter 2 for instance, but they cannot control city activities to the degree they were able to before globalization. It is this situation that allows the argument to be made that critical parts of city economies have been freed from the territoriality of states.

US cities in globalization

Popular studies of globalization, dismissed by Held *et al.* (1999) as 'hyper-globalist' because they underestimate the power of states, do sometimes have some interesting points to make. The leading German contribution to this genre by Martin and Schumann (1997) refers to contemporary times as 'the age of cities' (p. 18), but Martin and Schumann do not think all state sovereignty has disappeared: they identify the United States as 'the last nation which has still preserved a large degree of national sovereignty' (ibid.: 216). Not only the lone political superpower, having seen off the Japanese challenge in the 1990s it houses the most important economic segment of the world-economy. Thus according to Martin and Schumann, this particular state can 'set the rules for global integration' (ibid.: 216–17). This observation is intriguing on two counts. What can it mean for the world city network, specifically in terms of the supposed 'freedom' just discussed? And what can it in mean expressly for US cities in the world city network? The riddle is that globalization emerged from post-Second World War Americanization, and now it seems that the United States is somehow the only country that has managed to preserve a space of autonomy within globalization. Is such a circumstance discernible within the urban global analyses? The short answer is yes.

In the global urban analyses a 'global America' is briefly glimpsed in only three results: in New York joining with London as 'Main Street, World-Economy' at the centre of the configuration of urban arenas (Table 7.10 and Figure 7.4), in leading US cities as global service 'articulators' (Figures 6.4–6.7) and with US cities being well represented as command power centres (Figures 4.7–4.9). But these are the exceptions. In general, the various analyses reported above show that US cities are relatively separate from world cities in other countries. This includes even neighbouring Mexico and Canada: there are no analyses in which cities from the three North American countries appear together. But this aspect of world city network formation is much wider than eschewing neighbours. US cities consistently form a separate grouping or groupings of world cities. This is shown in the urban arenas beyond the centre (Tables 7.11–7.13); in the various experimental results using principal components analysis, where US cities stick closely together throughout; and in US city hinterworlds. In Chapter 5 only four US city hinterworlds were depicted; three of these were the only US cities with definable patterns of outside links: New York, Los Angeles and Miami. The tight-knit pattern of Indianapolis's hinterworld (Figure 5.8) represents to a large degree all other US city hinterworlds. In general, the global urban analyses show that US cities are always most strongly connected to other US cities. Less global, this is an 'exceptionist America' in the world city network.

Of course, the idea of an 'exceptionist America' is only a matter of degree. It must not be interpreted literally: in the global urban analysis data set every one of the 123 cities is connected to every other one. But there is no doubt that US cities appear to be less prominent in the world city network – that is, less well connected – than might be expected. This can be shown by comparing US world cities with European world cities in terms of their respective global network connectivities. For comparison, I focus on the European Union (EU) as an economic area that is roughly equivalent to the United States in economic size and wealth. The results are shown in Table 9.1, which includes

Table 9.1 *Global network connectivities: EU and US cities compared*

World rank	EU	EU rank	US rank	USA
1	London	1		
2			1	New York
4	Paris	2		
7			2	Chicago
8	Milan	3		
9			3	Los Angeles
11	Madrid	4		
12	Amsterdam	5		
14	Frankfurt	6		
15	Brussels	7		
17			4	San Francisco
25			5	Miami
27	Stockholm	8		
30	Dublin	9		
32	Barcelona	10		
33			6	Atlanta
37			7	Washington, DC
39	Vienna	11		
42	Lisbon	12		
44	Copenhagen	13		
48	Hamburg	14		
49	Munich	15		
50	Düsseldorf	16		
51	Berlin	17		
53	Rome	18		
56	Athens	19		
60			8	Boston
61			9	Dallas
62			10	Houston
63	Luxembourg	20		
68			11	Seattle
70	Helsinki	21		
73			12	Denver
74	Stuttgart	22		
75	Rotterdam	23		
76			13	Philadelphia
77			14	Minneapolis
81			15	St Louis
85			16	Detroit
92	Cologne	24		
93	Lyons	25		
96	Antwerp	26		
98			17	San Diego
101	Manchester	27		
105			18	Portland
106	Birmingham	28		
108			19	Charlotte
112			20	Cleveland
114			21	Indianapolis
119			22	Kansas City
120			23	Pittsburgh

twenty-eight EU and twenty-three US cities. The key point this table makes is not the slight additional numbers of EU cities but their general tendency to be ranked higher globally relative to equivalent US cities. The two capitals encapsulate this situation: both rank seventh in their respected economic areas but Brussels ranks fifteenth globally compared to Washington's thirty-seventh global ranking. This feature is consistent throughout the table: Barcelona and Houston ranked tenth locally, but thirty-second and sixty-second respectively globally; Luxembourg City and Cleveland are both twentieth locally, but sixty-third and 112th globally. The conclusion is inescapable: overall, US cities tend to be less globally connected than their European counterparts.

This is a very telling finding, but why should this be? There are two possible reasons for the United States' status as the leading economy within the world-economy not being translated into an equivalent roster of well-connected and incorporated world cities. First, there is what I call the 'shadow effect': non-US service firms developing a global strategy must locate in New York (and perhaps Chicago and Los Angeles) but decide on no further penetration of an already well-serviced US domestic market. They need a US presence for their non-US clients but do not wish to invest to service within the US market itself. Hence, for global services delivered by non-US firms, the concentration in New York casts a 'shadow', with fewer global services provided elsewhere in the country. Second, there is a 'comfort effect': leading US financial and business service firms may decide not to develop a global strategy and concentrate on expanding within the domestic market. Since this market is so large, such a strategy may be a superior one in being less risky and without the need to cope with expensive 'loss leader' foreign city offices. Certainly there is much less of an incentive for US financial and business service firms to embark on a transnational strategy than for firms in any other country.

What are the implications of this for the future? Clearly, the idea of a transnational 'freedom' for cities is less relevant in general to US cities, ensconced as they are within their state territory. Along with the pre-eminence of the US territorial state, it can be seen that the globalization versus state debate cannot be conducted as an abstract theoretical positioning. There is a world geography to this debate and it may well take a core–periphery type of pattern. As well as US cities remaining relatively territorial, it seems reasonable to argue that the cities likely to obtain most autonomy from states are to be found beyond the core, where states are weak and sometimes even failing. For instance, Rakodi (1998: 338) has shown that in the case of Kinshasa and Zaire, while the latter was collapsing politically, the city continued to function. Here the failing state had 'a very weak role' (ibid.) in a continuing urban administration. It can be surmised that cities and states in the rest of the world lie on a spectrum between the US extreme of territorialized cities and Africa's possible tendency towards more autonomous cities. But the power is concentrated at the territorial end; is this the beginning of a world politics in which the United States as superpower is the conservative defender of continuing the dominance of a space of places while the rest of the world moves towards increasing organization through a space of flows? With New York in the role of linking these two spaces?

The Dutch 'golden age' as economical politics

In presenting her 'utopian fantasy', Jacobs (1984: 217) introduces the concept of a 'pattern state' as a model for others. In the twentieth century the prime examples were the United States, the Soviet Union and the People's Republic of China. Clearly, Singapore has not been such a pattern state for city polities; rather than being emulated, it remains unique, and with the polity most similar, Hong Kong, experiencing the opposite of division – reintegration into China – the likelihood of Singapore leading a world city secessionist movement is not even on the horizon. Thus Jacobs can find no pattern state for her 'expedient multiplication of sovereignties' (ibid.: 218) argument, and so it is left in the realm of fantasy. However, I can find a historical example within the modern world-system that has potential for explicating the possibility of an increasing freedom for cities.

The idea of searching for a pattern state to illustrate Jacobs' (1984) utopia is problematic. What Jacobs is describing is a process, not a pattern. Furthermore, it is a process of multiple divisions, so that Singapore as a city-state from a single division is simply not relevant irrespective of its limited political influence. Jacobs (1984: 218) signals this when she corrects her use of 'pattern state' as a way of describing her secessions process by suggesting the alternative phrase 'pattern family-of-states' instead. Obviously, the end result of the secessions is to create a set of states for a given country, and therefore her new phrase appears to be very appropriate. However, it does not change the fact that no direct examples of the process are identified. I offer the seventeenth-century 'Dutch Republic' as a pattern family-of-states.

In world-systems analysis the 'Dutch Republic' (United Provinces) is normally identified as one of three hegemonic states that have defined the basic trajectory of the modern world-system (Wallerstein 1984). However, compared with the British in the nineteenth century and the Americans in the twentieth century, the seventeenth-century Dutch appear to be a pale shadow of what a 'world hegemon' should be. It hardly seems feasible that this new polity, a very small state both territorially and demographically, only a few decades old when it became hegemonic, could set the path along which the modern world-system embarked to eliminate all rival systems. Maurice Aymard (1982: 10) is the most explicit in problematizing the idea of the Dutch as hegemons: he asks whether the Dutch as the centre of a new world-economy created the last great city or the first great state.

Was the seventeenth-century 'Dutch Republic' a state?

Different authorities can be placed on either side of the 'state or city' debate. The city-position is that what was created in the seventeenth-century northern Netherlands was an Amsterdam city-state much in the manner of earlier city-states that dominated their world-economy such as Venice in its prime (e.g. Barbour 1963). Its proponents' main argument is based upon the quantitative economic dominance of Amsterdam in the northern Netherlands, which spilled over into the political sphere. The state-position is that the northern Netherlands constituted a new territorial state, taking its place in the inter-state system emerging at this time. The main argument for this view is that although the Dutch operated through a very distinctive state apparatus, it still was in fact a modern territorial state (t'Hart 1993). There is a third position: Braudel (1984), although initially favouring the city side, eventually comes down in the middle: the Dutch polity as a transition form between city-state and modern territorial state.

I will deal with the state option first since this is the dominant position, certainly within world-systems analysis, and it is one that I have subscribed to (Taylor 1996). The work of t'Hart (1989, 1993) carefully delineates the Dutch polity as a 'state of 58 cities' (1989: 666) that emerged in a war against Habsburg political centralization. Thus it is to be expected that the resulting polity would be a decentralized state, the opposite of the centralizing 'absolutist' states of the times. However, because the ideal model of the early modern state has come from the ranks of the latter, France to be precise, it follows that Dutch state credentials are left open to doubt. t'Hart's research strives to show the fiscal novelty of the Dutch in an alternative form of modern state making.

In effect, she sees the Dutch Republic as a victim of a particular European historiography that equates state with centralization. But, though breaking away from one orthodoxy, such arguments remain within even more deeply embedded ways of thinking. Let me explain by way of a parallel argument. In his famous discussion of the Dutch golden age, Schama (1987) has pointed out that the traditional interpretation of the late sixteenth-century Dutch rebellion as a nationalist revolt is merely a projection of nineteenth-century romantic historiography on to early modern Dutch politics. I would argue that the contemporary interpretation of the result of the Dutch rebellion as a modern state is a projection of twentieth-century social science statism on to seventeenth-century Dutch politics. The concept of the inter-state system leaves no geographical space for other than territorial organization of sovereign states, so its intellectual concomitant, embedded statism, leaves no theoretical space for other types of polity. Projected backwards, if the Dutch are to contribute to the making of the modern world, they could *only* have done this through possessing a modern state.

This logic is very clear in Israel's (1989) explanation for the Dutch inheriting Antwerp's economic network after its fall to Spanish troops in 1585. Initially the economic diaspora from Antwerp was widely dispersed, with 'north-western Germany' (primarily Hamburg) as the 'most likely candidate' to reap the benefits of Antwerp's demise (p. 33). But very soon the focus switched to the northern Netherlands because of the crucial need for the 'protection and active support of a powerful state' (ibid.: 36). Politically fragmented northern Germany simply could not compete. Hence the rise of Amsterdam in superseding Antwerp is due to the creation of a powerful state by the Dutch. This standard world-systems explanation equates state with defence, implying that the former is the only institution that can provide the latter. But if this conflation of state and defence is removed, it can be suggested that what the Dutch possessed was a highly defensible territory that was marshalled against attack by a geographically concentrated set of cities. Thus I can find no logical reason to continue with the orthodox state necessity argument: in the right circumstances, an alliance of cities should be able to organize successfully their own defence without the need to create a modern state.

In my view, constricted orthodox thinking has been aided by the posing of an alternative that can be shown to be flawed relatively easily. Quite simply, there was no 'Amsterdam city-state', for the fundamental reason that the city of Amsterdam did not rule this new political space either directly or indirectly by some subtle form of subterfuge that is difficult to reveal. All recent writers agree that the Dutch polity operated through shifting coalitions of cities, provinces and the stadholders (military governors). In key questions such as peace versus war, for all its economic importance, Amsterdam could be and was outmanoeuvred by counter-coalitions. Venice, a true city-state, could never have lost control of Venetian foreign policy! Hence if the choice of where to place the Dutch polity is between city-state and modern state, the latter almost wins by default. But this is only the case if the choice is limited to just these two possibilities.

Returning to t'Hart's 'state of 58 cities', removing reference to state leaves a political space of fifty-eight cities. Such a space is a polycentric city-region. This idea is picked up in some of the literature with references to, for example, a 'league of city-states' (Boogman 1979: 398) and 'a confederation of Venices' (Pocock 1992). It was in fact a unique political creation with no name – according to Schama (1987), the contemporary name 'United Provinces' was coined by an Englishman; 'Dutch Republic' is a more recent statist appellation. In reality it was an alliance of cities and provinces created by treaty in 1579. The resulting polycentric city-region polity had a very specific spatial structure: a vibrant core of many cities surrounded by a defensive frontier of fortifications (Taylor 1994). Braudel (1984: 202) calls the result a 'fortified island', in the middle of which is a 'high-voltage urban economy' (ibid.: 180). Of course, there had been polycentric city polities in the past – city leagues and trading networks as political institutions, such as the German Hanse described in Chapter 1, and coalitions of city-states to be found in northern Italy – but none of these formed a city-regional structure with a defensive shell in the manner of the Dutch. It is the latter's territorial defensive practices that make Braudel's argument for a halfway house between city and state seem so plausible. However, I prefer to interpret this political outcome as a distinctive polity, neither city nor state in the terms of this debate. Recent scholarship supports this position, most notably in the contrasts between Amsterdam and London in their respective 'golden ages' (O'Brien *et al.* 2001). The key point is that by locating the Dutch polity as a historical transition, Braudel eliminates its contemporary relevance. Thus instead of the Dutch polity being a stop on the evolutionary path to modern statehood, I will view it as a particular capitalist polity historically superseded by modern states but theoretically still available as a political category for future reference.

It is in this spirit that I consider the primary political dimensions of a polycentric city-regional polity. Such a polity will combine co-operation with competition among its constituents. In the case of the Dutch polity, co-operation is basic to the defensive function as written into the founding treaty of 1579 (for instance, joint responsibility for the fortifications – see original texts in Kossmann and Mellick (1974: 167–9)). However, it occurs also in the economic division of labour that developed. According to Braudel (1984: 182), Amsterdam was the apex of a pyramid of cities and co-operated with the following cast:

> industry prospered in Leyden, Haarlem and Delft; shipbuilding in Brill and Rotterdam; Dordrecht made a living from the heavy flow of traffic along the Rhine; Enkhuisen and Rotterdam controlled the fisheries of the North Sea; Rotterdam, again, the most important city after Amsterdam, handled the lion's share of trade with France and England; the Hague, the political capital . . .
>
> (Braudel 1984: 184)

But this division of labour was also a source of conflict within the polity. For instance, industry and trade interests differed with respect to war and peace, with Haarlem and Leiden leading the 'war party' that prevented the 'peace party', led by Amsterdam and Rotterdam, from concluding a truce in 1630 (Israel 1989: 56). Of course, all polities are a matrix of conflict and compromise, but what makes the Dutch polycentric city-region polity so unusual is that it had no institution to act as political umpire to resolve differences: a unanimity rule ruled. In modern states the 'umpire' role is the function of the executive constrained by the legislature and judiciary, backed up by the coercive arm of the state. The nearest the Dutch got to an executive with coercive capability was the stadholder position, which was periodically powerful but never attained sovereign status. This further suggests that we should not confuse the Dutch polity of the seventeenth century with a modern state. The Dutch political space was filled by a unique political entity, a polycentric city-region polity.

Political economy and economical politics

The Dutch 'golden age' coincides with the 'era of mercantilism'. Since contemporaries were forever complaining that the Dutch were 'ruled by merchants', it might be thought that the Dutch polity of the seventeenth century would be the prime example of mercantilism. Not so. The classic 'mercantilists' were the English and the French. So what about the Dutch? Traditionally not identified as mercantilist, in world-systems analysis they are brought to the very centre of the 'mercantile system' (Wallerstein 1980). Like the previous debate, this debate has an important historiographical dimension.

The term 'mercantile system' was defined by Adam Smith in the eighteenth century to describe the restrictive economic policies of states he was keen to critique. These were policies of partnership between economic elites and the monarch on the basis of the equation that more wealth equals more power in and for the state (Wilson 1958). However, in the late nineteenth century this pragmatic policy became reinterpreted as a policy of economic nationalism to justify state protectionism for new states (Coleman 1969). In this way, older states – England and especially France – became the 'pattern states' of mercantilism. The common missing link in these arguments was the target of these mercantilist policies. It was the Dutch who stimulated the need to devise restrictive policies: England and France were trying to counter the economic successes of the Dutch cities. Hence the so-called 'Dutch merchant state' was not mercantilist as defined by Smith since it was in its general interest, as hegemon, to have as few economic restrictions on trade as feasible (Wallerstein 1980; Taylor 1996). Of course, Smith's original definition of mercantilism, by omitting the Dutch, describes only half the world-economic process as it existed in the seventeenth century. If one uses a broader definition of mercantilism defined as political policy for the benefit of merchants, the Dutch become the super-mercantilists (Rowen 1978: 189). Without the need for partnership with the 'private interests' of a monarch, there was little or no dilution of 'public' economic policy except in emergencies when the stadholders played their military role.

As the discussion above shows, it is difficult to develop ideas on mercantilism without consideration of the state again. The 'super-mercantilist' conclusion merges with Boogman's (1978) thesis that the Dutch created a new economic-political *raison d'état* that contrasted with the standard political-military *raison d'état* of the times. Although very tempting – again I have used it in my work (Taylor 1996) – this idea contradicts the previous conclusion not to interpret the Dutch polity as a state; no state equals no *raison d'état*. As Braudel (1984: 205) tells it, 'for the Dutch, commercial interests effectively replaced *raison d'état*'. But this leaves no obvious way of characterizing the Dutch polity's pro-economic policies. To unravel this conundrum I return to Jacobs' heretical economics.

As is described in Chapter 2, at the core of Jacobsean economics is the revealing of a taken-for-granted assumption within macroeconomics that she terms the 'mercantilist tautology' (Jacobs 1984: 32). She argues that all subsequent schools of economic thought have shared the mercantilist purpose to construct policy levers to 'steer' the 'national economy'. In this sense they can all be termed 'political economy'; they are developments of economic ideas to serve the state. Thus statist mercantilism created a *political economy* discourse that has survived through to the embedded statism of the contemporary social science discipline of economics. The point is, of course, that the original statist mercantilism delineated only a 'half-system' component of the seventeenth-century world-economy: it missed out the Dutch, the target, the hegemon.

If the Dutch did not constitute a state, then their pro-merchant policies could not be described as political economy. Rather, I turn the term around and refer to Dutch policy as *economical* politics. The latter are, of course, policies that independent cities pursued before the rise of the modern state. In effect, what I am doing here is dividing Arrighi's (1994: 10) 'political capitalism' into two types: political economy and economical politics. The Dutch polity as a polycentric city-region continued city-centric economical politics while others created the political economy of statist mercantilism. The clearest indication of this difference can be seen in the work of Misra and Boswell (1997), where seventeenth-century Dutch leadership (as measured by ships) in trade is very much greater than in military seapower (from Modelski and Thompson 1988). This contrasts markedly with the great rival of the Dutch, England, where seapower was consistently much more important than trade. In other words, the Dutch were more of an economic power than a military one (i.e. economical politics), with England much more of a military power than an economic one (i.e. political economy).

Generally, economical politics can be seen in a concern for networks rather than territory. The Dutch polity of the seventeenth century was famously unconcerned with territorial expansion: as long as the frontier operated effectively as a defensive shield, no extra land was deemed necessary (Taylor 1996: 55). But this did not mean isolationism; far from it, for the Dutch were 'intimately concerned with the maintenance of favourable conditions in the world beyond their boundaries' (Wilson 1957: 6). And this meant developing and maintaining networks of trade and finance. Thus the Dutch were concerned to keep open the Sound to the Baltic and intervened in Swedish–Danish relations on numerous occasions to keep the seaway open. In stark contrast, the Dutch polity was adamant that the Scheldt should not be opened. In war it closed off Habsburg Antwerp from the sea, and the Dutch were just as keen to keep it closed in peacetime to stop this erstwhile world-city regaining any economic clout to rival Dutch cities. Hence economical politics does not translate simply into 'free trade'; rather, it is 'pro-trade' policy for the particular city or cities. This simply means the city or city-region attending to its networks, which was the basis of Dutch economical politics.

Such economical politics on a world-economy scale ended in 1672 with the demise of civilian leadership of the Dutch polity in the wake of the French invasion. It is from this point that the Dutch formed themselves into a state pursuing vigorous military and economic wars – political economy policies – against its rivals. As a state, the Dutch were a historical failure, but theoretically the closing off of the city-region polity option was of much more importance. Subsequent hegemons were explicitly statist, however liberal their policies: both the British Manchester liberals and Americans proclaiming that 'the business of America is business' can only be described as political economists. Thus the seventeenth-century Dutch polity of cities is to be intellectually treasured as a unique alternative to all-pervasive modern statism, especially given that it may have particular resonance for contemporary globalization.

Two routes to city success

It has been pointed out on many occasions that while the rest of Europe was suffering its 'crisis of the seventeenth century', the Dutch bucked this trend with their 'golden age'. In short, the Dutch are interpreted as having the only successful state of their era. However, from a city-centric position there is a different conclusion. Certainly the cities of the Dutch were very successful in the seventeenth century, and this economic vibrancy was reflected in their rapidly growing populations. But they were not the only cities to experience rapid demographic growth: Madrid, Vienna, Paris and London were also cities that prospered in the seventeenth century. Of course, the latter quartet grew for different reasons as compared with the Dutch cities: they were the beneficiaries of political restructuring that produced the more centralized state-form that the Dutch rebelled against. The growth of these capital cities can be said to reflect a new political vibrancy. Thus early modern Europe shows two distinct urban routes to success: as well as an economical politics route taken by the Dutch, there was a political economy route within the emerging inter-state system.

Which route is taken has very profound inter-city consequences. The effects of political economy processes on inter-city relations is starkly illustrated by the decline of the cities of central Castile in the early seventeenth century (Ringrose 1989; Albaladejo 1994). Using Jacobs' terminology, this is a case of 'city killing' *par excellence*. In the sixteenth century central Castile had a successful urban hierarchy of cities with Toledo at the top. This lasted until the new Castilian capital city, Madrid, rapidly expanded to

Table 9.2 *Populations (000s) in selected Castilian and Dutch cities, 1570–1670*

	c. *1570*	c. *1600*	c. *1630*	c. *1670*
Toledo	65	65	20	15
Madrid	35	65	175	120
Amsterdam	30	60	116	200
Leiden	15	26	54	72

Sources: Ringrose (1989) and Israel (1995)

outgrow Toledo and all the other local cities put together. The effect of Madrid's growth was catastrophic. Instead of representing new economic investment in the region, it acted as a capital-sink, politically distorting local markets and thus destroying other city economies. The pertinent figures are shown in Table 9.2, where the two leading cities of central Castile are compared to the two largest cities of central Holland. Note that initially Madrid, on its political economy trajectory, actually outgrows Amsterdam. However, the real contrast is between the fate of the two 'second cities': unlike Toledo's decline, Leiden's growth actually keeps up with that of its illustrious local rival. This stark difference between Toledo and Leiden represents a key dissimilarity between political economy and economical politics. While the rise of Madrid as a great Habsburg capital city was destroying an adjacent urban economic structure, the rise of Amsterdam was accompanied by the rise of other Dutch cities in a mutually reinforcing polycentric city-region. This is the 'division of labour' described by Braudel (1984: 182) and noted above. Such inter-city mutuality is central to economical politics.

In conclusion: it is political economy that kills cities (other than capital cities), whereas cities prosper under conditions of economical politics. Madrid and Toledo were playing out a zero-sum game of competition, while Amsterdam and Leiden were in a 'win–win' situation. A question that arises for cities in globalization is, therefore, whether the leading cities such as London and New York affect other cities like a modern-day Habsburg Madrid or like a modern-day 'golden-age' Amsterdam. However, my purpose is not to 'learn lessons' from the golden-age Dutch by drawing simple analogies across the centuries. This very special polity can never be a 'pattern family-of-cities' for twenty-first-century cities in globalization, but it can provide some rare network food for thought.

Beyond political economy

The basic value of the Dutch multi-nodal city polity is that it provides a historical alternative to political economy within the modern world-system. Political economy as 'state economics' dominates thinking about change in the modern world, but it does not have to be that way. This is important for considering future change because political economy is a 'mosaic discourse' that promotes a space of places in a world in which the salience of spaces of flows is growing. Following my supposition that globalization and the world city network are not bundles of processes that necessarily promote capital, in this brief conclusion I broach two research agendas that explore beyond political economy.

Competition and co-operation in the world city network

Although it is not expected that political economy will disappear with globalization, it is legitimate to ask whether there are indications of a new economical politics appearing.

Political economy is reflected globally in an international relations (IR) that is essentially 'realist' in nature, a thought and practice that is about competition between states. There is an 'idealist' counter to this dominant school of IR, but in practice co-operation across states always fails when perceived threats to 'national interest' are involved: IR defines an aggressive world. This competitive ethos in social change is to be found in the urban literature as city competition (see, for example, the survey by Kresl (1995) and the papers brought together by Lever and Turok (1999)). Operating as a sort of 'lower-level IR', traditional city development studies emphasize competition through city hierarchies. This is, of course, the assumption behind Petrella's dystopia, the spectre that haunts all world city studies. However, the IR of states operates in a mosaic world of boundaries, whereas cities do not form mosaic spaces, they are nodes in spaces of flows. The critical point is that for cities in networks it is co-operation that is inherent in inter-city relations. In general, mutuality is at the heart of all networks; without a coincidence of interests, networks simply collapse. In the case of the world city network the mutuality exists through the service firms and their global location strategies. Investing in expensive offices across many cities means that these firms have a vested interest in all the cities they operate through. I proceed on the basis that the prime reality of contemporary inter-city relations is a symbiosis among world cities as a global network.

This conclusion provides a basis for searching out economical politics in contemporary globalization (Beaverstock et al. 2001). London and Frankfurt are a particularly interesting pair of world cities because the location of the European Central Bank in Frankfurt, followed by the launch of the euro currency, led to much speculation in the financial press that Frankfurt was 'catching up' with London. In other words, relations between London and Frankfurt were enframed as competition. A review of the press in both cities found this to be by far the dominant discourse, a sort of lower-level political economy competition between Germany and the United Kingdom. However, a very different story emerges when service practitioners in both cities are interviewed. The idea that London and Frankfurt are in competition is an alien notion to their economic practices. Working with colleagues in their firm's office in the other city, and visiting that office in the other city, meant that the two cities were commonly seen as part of a single practice. There was a division of labour – London dealing with more global projects, Frankfurt with European projects – and the dominating process was one of co-operation. This idea was even found in the interviews with urban officials working to promote city institutions in both places. Quite simply, the view was that 'what's good for London is good for Frankfurt and vice versa'. Growth in spheres of work attended to in Frankfurt has potential for spilling over into London's global practice, and the latter provides opportunities for developing Frankfurt's European work. In short, London–Frankfurt is an important European dyad in a world city network where co-operative relations interlock the cities.

What about the large literature on competition between cities that counters the network mutuality argument? Such research advises urban policy makers to mould their locales in order to attract investment in what is apparently deemed to be a zero-sum game. However, even in this literature, problems are found with the 'competition model' (Lever 1999: 1030) and there is recognition of mutuality: 'Plainly cities compete. . . . Equally, cities cooperate' (Begg 1999: 807). This seems like a recipe for policy confusion: how to compete and co-operate at the same time? Perhaps city competition is not quite so 'plain'. Budd (1998) follows Krugman (1994) to argue that it is only firms that compete (in markets), not 'nations'; Budd thinks this critique of 'non-market competition' applies 'doubly' to cities (1998: 670). This is consistent with the approach adopted in this book: by conceptualizing the world city network as an interlocking network, the competition is between global service firms in world markets for financial and business services; it is not between the cities. This idea is basic to avoiding reification of cities in Chapter 3. The policy logic of this is for city authorities to attend to much more than just their place; they need to devote as much political energy to their networks (Beaverstock *et al.* 2001) – that is, other places beyond their specific jurisdictions. It is not at all clear what form such political practice – an economical politics – might take, except that it will need to be a new network politics beyond the usual place-based politics. However, I am persuaded to make the one future prediction of this chapter: big city mayors (increasingly independent of national political parties) will be important contributors to the creation of a global economical politics for the twenty-first century (see Bagnasco and Le Gales 2000: 27–8). But how can such a network politics be democratic?

Transnational democracy?

There is a contradiction at the heart of current democratic theory and practice: just as representative democracy has spread to more countries than ever before, its supporters have massive new doubts as to its effectiveness in defending the interests of the electorate. Giddens (1999: 71–2) calls this a 'democratic paradox' in contemporary globalization: clearly defined and accountable government seems to be giving way to a nebulous governance – who ever voted for a governance? Islin (2000) describes the challenges this poses for modern citizenship.

The fundamental problem for democracy in globalization is the erosion of a franchised people equating to a 'community of fate' in the sense that they can control their own future (Held *et al.* 1999: 30–1, and see, in particular, figure 10.1, p. 224). Modern democracy is intrinsically bound up with the nation-state. The contemporary state has been legitimated as nation-state where the 'nation' is an imagined community (Anderson 1983). The political economy democracy model incorporates this designated 'people' with a common past as a political actor whose elected leaders are expected to achieve continuous economic 'national development' into the future. It is this combining of past and future – roots with fate – in the nation that makes democracy a plausible collective enterprise: the imagined community is also a common community of fate. However, the basic political threat of globalization is the separation of these two idealized communities as a result of economic forces beyond the state becoming increasingly dominant. Thus 'rule by the people' is meaningless if their elected leaders are deprived of the 'levers of power'.

Modern democracy is territorial; it is a bounded practice. In Castells' (1996) terms, this means that it is condemned to operate in space of places, leaving the space of flows to other activities (economics/work, culture/identity). This is the reason why it is so vulnerable in a globalizing world. There is a simple logical solution: territorial organization of politics has to give way to a non-territorial democratic form. But actual politics cannot be so simple; where is the historical process through which politics can be restructured into a network model? I am sceptical of the possibilities of creating a new democratic 'non-national community'. Certainly there is the technological potential for an electronic democracy, but this individualizes political activity. This may work for single-issue politics but it is not at all clear how a sense of community could be developed, however sophisticated virtual alternatives to real face-to-face communication become. Such 'online democracy' in a virtual space would have difficulties in building allegiances and defining communities of fate. This is democracy without demos, better termed 'ego-ocracy'. A modern democracy for a complex and ever-changing world requires an institutional anchor.

There are two non-territorial institutions that have become prominent in discussions of 'global governance': non-governmental organizations (NGOs) and social movements. While very interesting for their promotions of new participation in political decision making, both NGOs and movements have their own 'democratic deficits' (Princen and Finger 1994: 12). When losing territorial referents these institutions seem also to lose community attachment – who do these institutions represent beyond the self-selection of their own network of members? At least bounded democracy has provided representation for, and allowed participation by, all people within its territory.

An alternative approach is to think in terms of both place and flows simultaneously. The argument is straightforward: spaces of places are necessary for building an imagined community, while spaces of flows are necessary for any meaningful depiction of a community of fate. Cities (or city-regions) and their networks are the obvious geographical candidates that fit both needs. While being nodes within urban networks, cities are also places with long histories of being distinctive communities; as noted previously, most cities are older than the states that govern them. The key point is that cities, as the crossroads of society, are inherently non-territorialist. If globalization is in the process of 'unbounding' them, then the full network potential of cities beyond the nation-state can be made available for creating a new economical politics. The political raw material is already there: world cities are recognized as immensely polarized communities, they are where global capital meets democracy.

Ideas of citizenship and democracy originated in cities, and the question being asked is whether they can be reconstituted there (Douglas and Friedmann 1998; Islin 2000; Friedmann 2001). Cities already constitute 'communities', so constructing them as critical political units of imagination is feasible, but what of their possible role as communities of fate? Under conditions of globalization it is no more possible to create communities of fate from cities than from states. However, because of their inherent mutuality, it would be easier to construct collectives of cities as communities of fate than to combine states. Furthermore, the 'transnationalism' of state collectives, such as the European Union, remains territorial, albeit at a larger scale. This means that democracy in such institutions remains bounded in a space of places. In contrast, city collectives – new city leagues – are networks where republicans can finally eschew their modern penchant for mimicking territorial kingdoms and return to city networks reflecting specific mutual interests. For instance, Manchester and Lyons may come to see that they have more in common with each other than they have with their respective capital cities, London and Paris. For both 'second' cities, the experience of being in the shadow of major global cities might provide a basis for identifying a new community of fate with other cities similarly afflicted. Such leagues are currently rare in a boundary-obsessed political economy world, but there is one example that is instructive because it contrasts directly with territorial organization. In the north-west Mediterranean there are two rival cross-border political alliances: a 'Euro-region' combining Catalonia, Languedoc-Roussillon and Midi-Pyrenees, and the 'C6 network' consisting of Barcelona, Montpellier, Palma de Mallorca, Toulouse, Valencia and Zaragoza (Morata 1997). Modest in scale, nevertheless these two opposing organizations represent two different political worlds – territorial and network – with different potentials for transnational democracy under conditions of increasing globalization.

Processes require agents: who will be the instrument for relocating the demos from territorial nation to city network? Put crudely: who will do the deed, who are the putative agents for creating transnational democratic forms? The story of producing national democratic forms in the West was a reluctant bourgeoisie being forced to concede more and more radical demands until universal suffrage was achieved. In the process the original (pre-democratic) 'cadre parties' coached the new radical mobilizing parties into being 'good party citizens' on the promise that their turn in government would come (Taylor 1999: 82–4). The resulting integral states each with their own 'demos' are what globalization is challenging. It seems highly unlikely in the future that this particular configuration of political forces can be replicated outside state confines to produce a transnational democracy.

The changing nature of what were the individual cadres of the system is a major obstacle to such replication. Traditionally located between capital and labour, these 'brain workers', notably in the 'professions', took leadership roles in pressing for political reform. In the informational age, however, the new knowledge workers are being transmuted into something that looks very much like 'knowledge capitalists'. For instance, large share options have become the norm for key workers in some sectors (e.g. in finance), whereas in other sectors practitioners are 'partners' in what are in reality multinational firms (e.g. in global law). Always in an ambiguous position as 'professionals', they are managed through codes of service to clients in their respective sectors, while at the same time operating as profit-making firms and partnerships in the marketplace. The key point is that globalization's network society is changing the balance between service and profit to the detriment of the former. Products, in their design and implementation, are embodied in the knowledge of these professionals, who are among the most successful groups of 'capitalists' in the world today. It is the thesis of this book that they are the prime creators of the world city network, which is their global workplace.

These 'ex-cadres', now 'knowledge capitalists', are politically interesting for two reasons. First, their global interests will clash with national interests, and when this happens they can be mobilized against the state. Their interests coincide with those of London, not the United Kingdom, with those of New York and not the United States, and so on. Second, because they are located in all the major cities they cannot be mobilized in practices for inter-city competition. They are emphatically not local capitalists; they are 'network capitalists', and it is in their interests to see London *and* Frankfurt prosper. 'Competition between cities' leading to inducements offered by city governments will be important to them – it is something they can take advantage of to enhance profits – but they will have no inherent reason to take sides. Service-city 'capitalists' should be major supporters of co-operation between cities – promoters, perhaps, of city leagues.

Of course, like the national bourgeoisies of the past, this new network bourgeoisie will have no direct interest in promoting a democratic politics. Its members constitute a global plutocracy (van der Pijl 2002), directly reflected in world cities through their high levels of economic polarization (Sklair 1998). It is an understatement to say that this is not good raw material for creating a new demos. But the twenty-first-century city network is not necessarily worse than the nineteenth-century nation-state as a democratic nursery. The latter requires Falk's (2000) 'globalization from below' to create a new progressive politics. This will have to involve a range of new network processes through NGOs and new social movements and their ilk, as mentioned previously (Friedmann 2001). The key addition that cities provide is an alternative community grounding for such a politics – for instance, in the role of strong city mayors mediating between the global and the local. To be effective, the latter has to be radicalized, not just in one city, but as part of a new global wave of radicalization. Castells (1999: 302) calls this 'grassrooting the space of flows'. In this way the current global hegemony of neo-liberal market ideology is being challenged through reasserting that cities are for citizens (Douglas and Friedmann 1998). There is no reason to believe that with a different phase of the world-economy, alliances between local-city political parties and networked NGOs and social movements can change things around; certainly, historically radicalism has always been very cyclical in nature (Silver 1995). It may not be long before it is time to reassess the investment of radical political effort that goes into territorial states and release some of this energy for experimenting with building new network 'demos-es' for a progressive economical politics.

Appendix A: Global service firms (the 'GaWC 100')

Accountancy

1 Ernst & Young
2 Arthur Andersen
3 MacIntyre Strater International (MSI)
4 IGAF
5 AGN Network
6 BDO
7 Grant Thornton International
8 Horwath International
9 KPMG
10 Summit International + Baker Tilly
11 RSMi
12 Moores Rowland International
13 HLB International
14 Moore Stephens International Network
15 Nexia International
16 PKF International Association
17 Fiducial International
18 PricewaterhouseCoopers

Advertising

19 Impiric
20 TMP
21 Hakuhodo
22 Draft Worldwide
23 Densu Young and Rubicam + Young and Rubicam
24 D'Arcy
25 FCB
26 Saatchi and Saatchi
27 Ogilvy
28 BBDO Network
29 McCann-Erickson WorldGroup
30 J. Walter Thompson
31 Euro RSC6
32 CMG (Carlson Marketing Group)
33 Asatsu DK

Banking/finance

34 WestLB (Westdeutsche Landesbank Gironzentrale)
35 Dresdner Bank
36 Commerzbank
37 Deutsche Bank
38 Chase
39 BNP Paribus
40 ABN-AMRO
41 Rabobank International
42 UBS
43 ING
44 Barclays
45 Fuji Bank
46 Bayerische HypoVereinsbank
47 Bayerische Landesbank Girozentral
48 Sakura Bank
49 Sumitomo Bank
50 Sanwa
51 J. P. Morgan
52 BTM (Bank of Tokyo-Mitsubishi)
53 DKB (Dai-Ichi Kangyo Bank)
54 HSBC
55 Citibank
56 Credit Suisse/First Boston

Insurance

57 Allianz Group
58 Skandia Group
59 Chubb Group
60 Prudential
61 Reliance Group Holdings
62 Winterthur
63 Fortis
64 CGNU
65 Liberty Mutual
66 Royal & SunAlliance
67 Lloyd's

Law

68 Latham & Watkins
69 Morgan Lewis
70 Baker & McKenzie
71 Clifford Chance
72 Jones Day
73 Freshfields Bruickhaus Deringer
74 Allen & Overy

75 Dorsey & Whitney
76 Linklaters – Alliance
77 White & Case
78 Cameron McKenna
79 Morrison & Foerster
80 Lovells
81 Skadden, Arps, Slate, Measher, & Flom
82 Sidley & Austin
83 Coudert Brothers

Management consultancy

84 Towers Perrin
85 Logica Consulting
86 Watson Wyatt
87 Sema Group
88 CSC
89 Hewitt Associates
90 IBM Worldwide
91 Mercer Management Consulting
92 Boston Consulting Group
93 Deloitte Touche Tohmatsu
94 Booz Allen & Hamilton
95 A. T. Kearney
96 McKinsey
97 Bain & Company
98 Compass
99 Andersen Consulting
100 Gemini Consulting/Cap Gemini (Ernst & Young)

Appendix B: List of cities

London	Warsaw	Perth
New York	Seoul	Lima
Hong Kong	Lisbon	St Louis
Paris	Johannesburg	Bangalore
Tokyo	Copenhagen	Bucharest
Singapore	Budapest	Karachi
Chicago	Manila	Detroit
Milan	Montreal	Wellington
Los Angeles	Hamburg	Calcutta
Toronto	Munich	Ho Chi Minh City
Madrid	Düsseldorf	Manama
Amsterdam	Berlin	Jeddah
Sydney	New Delhi	Tel Aviv
Frankfurt	Rome	Cologne
Brussels	Dubai	Lyons
São Paulo	Bogotá	Cape Town
San Francisco	Athens	Riyadh
Mexico City	Santiago	Antwerp
Zurich	Caracas	Adelaide
Taipei	Cairo	San Diego
Mumbai	Boston	Nairobi
Jakarta	Dallas	Quito
Buenos Aires	Houston	Manchester
Melbourne	Luxembourg City	Chennai
Miami	Beirut	Hamilton (Bermuda)
Kuala Lumpur	Vancouver	Calgary
Stockholm	Oslo	Portland
Bangkok	Geneva	Nassau
Prague	Seattle	Birmingham
Dublin	Rio de Janeiro	Charlotte
Shanghai	Helsinki	Guangzhou
Barcelona	Montevideo	Casablanca
Atlanta	Brisbane	Port Louis
Moscow	Denver	Cleveland
Istanbul	Stuttgart	Bratislava
Beijing	Rotterdam	Indianapolis
Washington, DC	Philadelphia	Abu Dubai
Auckland	Minneapolis	Kiev
Vienna	Panama City	Kuwait

Nicosia	Belo Horizonte	Banda SB
Kansas City	Windhoek	Doula
Pittsburgh	Palo Alto	Salvador
Sofia	Lille	Omaha
Zagreb	La Paz	Gabeorone
Lagos	Kampala	Port of Spain
Amman	Hartford	Managua
Guayaquil	Göteborg	Berne
Ruwi	Tallinn	Tashkent
Osaka	Doha	Hyderabad
Monterrey	Richmond	Yokohama
Bilbao	Vilnius	Tijuana
Guatemala	Buffalo	Essen
Abidjan	Kingston	Norwich
Valencia	Bordeaux	Dalian
Harare	Christchurch	Brasilia
Asunción	Honolulu	Nagoya
Bristol	Ljubljana	Luanda
Baltimore	Belfast	Grenoble
Leeds	Edmonton	Belgrade
Glasgow	Curitaba	Pretoria
San José (CR)	Limassol	Naples
Marseilles	Nottingham	Bergen
Phoenix	Turin	Penang
Tunis	Winnipeg	Quebec
Almaty	Tegucigalpa	Sheffield
St Petersburg	Ottawa	Port Moseley
Edinburgh	Dar es Salaam	Bonn
Colombo	Basle	Reykjavik
Hanoi	Las Vegas	Cardiff
Hobart	Nuremberg	Yangon
Cincinnati	Shenzen	Aarhus
Accra	Seville	Macao
Santo Domingo	Maputo	Kyoto
Dhaka	Tehran	Suva
Tampa	Malmö	Genoa
San Salvador	Utrecht	Mainz
Riga	Dakar	Georgetown
Lusaka	Newcastle	Ahmadabad
Lahore	Liverpool	Tianjin
Dresden	Medellín	Ciudad Juárez
Columbus	New Orleans	Recife
Strasbourg	Baku	Addis Ababa
San Jose (CA)	Hanover	Dortmund
Leipzig	Bologna	Bangung
Rochester	Aberdeen	Kobe
Islamabad	Canberra	Bulawayo
Labuan	Lausanne	Pusan
Durban	Sacramento	Plymouth
Porto Alegro	Southampton	Damascus
Guadalajara	The Hague	Alexandria

Wilmington
Rabat
Palermo
Mannheim
Ankara
Linz
Tirana
Kinshasa
Mombasa
Medan
Sanaa
Algiers
Jerusalem
Freetown
Trieste

Sarajevo
Minsk
Port au Prince
Yerevan
Lomé
Tripoli
Tblisi
Liège
Xiamen
Batam
Cracow
Khartoum
Nanjing
Malacca
Venice

Manaus
Havana
Kawasaki
Yaoundé
Jaipur
Monrovia
Ulan Bator
Rawalpindi
Conakry
Djibouti
Baghdad
Kabul
Brazzaville
Lucknow
Pyongyang

 **Bibliography**

Abbott, C. (1993) 'Through flight to Tokyo: sunbelt cities and the new world economy, 1960–1990', in A. R. Hirsch and R. A. Mohl (eds) *Urban Policy in Twentieth Century America*, New Brunswick, NJ: Rutgers University Press.

Abbott, C. (1996) 'The internationalisation of Washington, D.C.', *Urban Affairs Review*, 31: 571–94.

Abbott, C. (1999) *Political Terrain: Washington, D.C. from Tidewater Town to Global Metropolis*, Chapel Hill: University of North Carolina Press.

Abrams, P. (1978) 'Towns and economic growth: some theories and problems', in P. Abrams and E. A. Wrigley (eds) *Towns in Society*, Cambridge: Cambridge University Press.

Abu-Lughod, J. L. (1989) *Before European Hegemony: The World System, AD 1250–1350*, New York: Oxford University Press.

Abu-Lughod, J. L. (1999) *New York, Los Angeles, Chicago: America's Global Cities*, Minneapolis: University of Minnesota Press.

Agnew, J. (1993) 'The territorial trap', *Review of International Political Economy*, 1: 53–80.

Agnew, J. (1997) 'Commentary and criticism: fifty years after the publication of Harris and Ullman's "The nature of cities"', *Urban Geography*, 18: 4–6.

Aharoni, Y. (1993a) 'Globalization of professional business services', in Y. Aharoni (ed.) *Coalitions and Competition*, London: Routledge.

Aharoni, Y. (ed.) (1993b) *Coalitions and Competition*, London: Routledge.

Albaladejo, P. F. (1994) 'Cities and the state in Spain', in C. Tilly and W. P. Blockmans (eds) *Cities and the Rise of States in Europe, AD 1000 to 1800*, Boulder, CO: Westview.

Allen, J. (1997) 'Economies of power and space', in R. Lee and J. Willis (eds) *Geographies of Economies*, London: Arnold.

Allen, J. (1999) 'Cities of power and influence: settled formations', in J. Allen, D. Massey and M. Pryke (eds) *Unsettling Cities*, London: Routledge.

Amin, A. and Graham, S. (1999) 'Cities of connection and disconnection', in J. Allen, D. Massey and M. Pryke (eds) *Unsettling Cities*, London: Routledge.

Amin, A. and Thrift, N. (2002) *Cities: Reimagining the Urban*, Cambridge: Polity Press.

Anderson, B. (1983) *Imagined Communities*. London: Verso.

Andersson, A. E. (2000a) 'Gateway regions of the world: an introduction', in A. E. Andersson and D. E. Andersson (eds) *Gateways to the Global Economy*, Cheltenham, UK: Edward Elgar.

Andersson, A. E. (2000b) 'Financial gateways', in A. E. Andersson and D. E. Andersson (eds) *Gateways to the Global Economy*, Cheltenham UK: Edward Elgar.

Andersson, A. E. and Andersson, D. E. (eds) (2000) *Gateways to the Global Economy*, Cheltenham, UK: Edward Elgar.

Andersson, D. E. (2000) 'The role of institutions and self-organizing networks in the economic history of regions', in A. E. Andersson and D. E. Andersson (eds) *Gateways to the Global Economy*, Cheltenham, UK: Edward Elgar.

Arrighi, G. (1994) *The Long Twentieth Century*, London: Verso.

Aymard, M. (1982) 'Introduction', in M. Aymard (ed.) *Dutch Capitalism and the European World-Economy*', Cambridge: Cambridge University Press.

Bagchi-Sen, S. and Sen, J. (1997) 'The current state of knowledge in international business in producer services', *Environment and Planning A*, 29: 1153–74.

Bagnasco, A. and Le Gales, P. (2000) 'Introduction: European cities – local societies and collective actors?', in A. Bagnasco and P. Le Gales (eds) *Cities in Contemporary Europe*, Cambridge: Cambridge University Press.

Barbour, V. (1963) *Capitalism in Amsterdam in the Seventeenth Century*, Ann Arbor: University of Michigan Press.

Barnett, R. J. and Muller, R. E. (1974) *Global Reach*, New York: Simon & Schuster.

Bartlett, C. J. (1984) *The Global Conflict, 1880–1970*, London: Longman.

Bassett, K. and Short, J. (1989) 'Development and diversity in urban geography', in D. Gregory and R. Walford (eds) *Horizons in Human Geography*, London: Macmillan.

Beaverstock, J. V. (1991) 'Skilled international migration: an analysis of the geography of international secondments within large accountancy firms', *Environment and Planning A*, 23: 1133–46.

Beaverstock, J. V., Smith, R. G. and Taylor, P. J. (1999a) 'A roster of world cities', *Cities*, 16: 445–58.

Beaverstock, J. V., Smith, R. G. and Taylor, P. J. (1999b) 'The long arm of the law: London's law firms in a globalizing world economy', *Environment and Planning A*, 31: 187–92.

Beaverstock, J. V., Smith, R. G. and Taylor, P. J. (2000a) 'World city network: a new meta-geography?', *Annals of the Association of American Geographers*, 90: 123–34.

Beaverstock, J. V., Smith, R. G. and Taylor, P. J. (2000b) 'Geographies of globalization: United States law firms in world cities', *Urban Geography*, 21: 95–120.

Beaverstock, J. V., Hoyler, M., Pain, K. and Taylor, P. J. (2001) *Comparing London and Frankfurt as World Cities: A Relational Study of Contemporary Urban Change*, London: Anglo-German Foundation for the Study of Industrial Society.

Begg, I. (1999) 'Cities and competitiveness', *Urban Studies*, 36: 795–809.

Berman, M. (1988) *All That Is Solid Melts into Air*, New York: Penguin.

Berry, B. J. L. (1961) 'City size distributions and economic development', *Economic Development and Cultural Change*, 9: 573–88.

Berry, B. J. L. (1964) 'Cities as systems within systems of cities', *Papers of the Regional Science Association*, 13: 147–63.

Berry, B. J. L. and Garrison, W. L. (1958) 'Alternative explanations of rank-size relationships', *Annals of the Association of American Geographers*, 48: 83–91.

Berry, B. J. L. and Horton, F. E. (1970) *Geographic Perspectives on Urban Systems*, Englewood Cliffs, NJ: Prentice Hall.

Berry, B. J. L. and Meltzer, J. (1967) *Goals for Urban America*, Englewood Cliffs, NJ: Prentice Hall.

Birdsall, S. S. (1980) 'Alternative prospects for America's urban future', in S. D. Brunn and J. O. Wheeler (eds) *The American Metropolitan System: Present and Future*, New York: Wiley.

Blowers, A., Hamnett, C. and Sarre, P. (eds) (1974) *The Future of Cities*, London: Hutchinson.

Board, C., Davies, R. J. and Fair, T. J. D. (1978) 'The structure of the South African space economy', in L. S. Bourne and J. W. Simmons (eds) *Systems of Cities*, New York: Oxford University Press.

Boogman, J. C. (1978) 'The *raison d'état* politician Johan de Witt, *Low Countries History Yearbook*: 55–78.

Boogman, J. C. (1979) 'The Union of Utrecht: its genesis and consequences', *Bijdragen Mededlingen Detreffende de Geschiedenis der Nederlanden* 94: 277–407.

Borchert, J. R. (1967) 'American metropolitan evolution', *Geographical Review*, 57: 301–32.

Boulding, K. E. (1978) 'The city as an element in the international system', in L. S. Bourne and J. W. Simmons (eds) *Systems of Cities*, New York: Oxford University Press.

Bourne, L. S. (1975) *Urban Systems: Strategies for Regulation*, Oxford: Clarendon Press.

Bourne, L. S. and Simmons, J. W. (eds) (1978) *Systems of Cities*, New York: Oxford University Press.

Braudel, F. (1984) *The Perspective of the World*, London: Collins.

Brooker-Gross, S. R. (1980) 'Uses of communication technology and urban growth', in S. D. Brunn and J. O. Wheeler (eds) *The American Metropolitan System: Present and Future*, New York: Wiley.

Brotchie, J., Batty, M., Blakely, E., Hall, P. and Newton, P. (1995) *Cities in Competition*, Melbourne: Longman.

Brown, E. D., Catalano, G. and Taylor, P. J. (2002) 'Beyond world cities: Central America in a global space of flows', *Area*, 34: 139–48.

Brown, L. R. (1973) *World Without Borders*, New York: Vintage.

Brunn, S. D. and Wheeler, J. O. (1980) *The American Metropolitan System: Present and Future*, New York: Wiley.

Budd, L. (1998) 'Territorial competition and globalisation: Scylla and Charybdis of European cities', *Urban Studies*, 35: 663–86.

Bunge, W. (1988) *Nuclear War Atlas*, Oxford: Blackwell.

Burt, R. S. (1983) *Corporate Profits and Cooptation*, New York: Academic Press.

Camagni, R. P. (1993) 'From city hierachy to city network: reflections about an emerging paradigm', in T. R. Lakshmanan and P. Nijkamp (eds) *Structure and Change in the Space Economy*, Berlin: Springer-Verlag.

Camagni, R. P. (2001) 'The economic role and spatial contradictions of global city-regions', in A. J. Scott (ed.) *Global City-Regions*, New York: Oxford University Press.

Cappelin, R. (1991) 'International networks of cities', in R. Camagni (ed.) *Innovation Networks: Spatial Perspectives*, London: Belhaven.

Castells, M. (1977) *The Urban Question*, London: Arnold.

Castells, M. (1989) *The Informational City: Information Technology, Economic Restructuring and the Urban-Regional Process*, Oxford: Blackwell.

Castells, M. (1996) *The Rise of Network Society*, Oxford: Blackwell.

Castells, M. (1999) 'Grassrooting the space of flows', *Urban Geography*, 20: 294–302.

Castells, M. (2001) *The Rise of Network Society*, 2nd edn, Oxford: Blackwell.

Chisholm, M. (1967) 'General systems theory and geography', *Transactions of the Institute of British Geographers*, 42: 45–52.

Cliff, A. D., Haggett, P., Smallman-Raynor, M. R., Stroup, D. F. and Williamson, G. D. (1995) 'The application of multidimensional scaling methods in epistemiological data', *Statistical Methods in Medical Research*, 4: 102–23.

Coffey, W. J. (2002) 'The geography of producer services', *Urban Geography*, 21: 170–83.

Cohen, R. B. (1981) 'The new international division of labor, multinational corporations and urban hierarchy', in M. Dear and A. Scott (eds) *Urbanization and Urban Planning in Capitalist Society*, London: Methuen.

Coleman, D. C. (ed.) (1969) *Revisions in Mercantilism*, London: Methuen.

Cook, T. (1984) 'A reconstruction of the world: George R. Parkin's British Empire map of 1893', *Cartographia*, 21: 53–65.

Corbridge, S., Martin, R. and Thrift, N. (eds) (1994) *Money, Power, Space*, Oxford: Blackwell.

Cox, K. R. (1997) 'Introduction: globalization and its politics in question', in K. R. Cox (ed.) *Spaces of Globalization*, New York: Guilford.

Crahan, M. E. and Vourvoulias-Bush, A. (eds) (1997) *The City and the World: New York's Global Future*, New York: Council of Foreign Relations.

Daniels, P. W. (1985) *Service Industries: A Geographical Appraisal*, London: Methuen.

Daniels, P. W. (1993) *Service Industries in the World Economy*, Oxford: Blackwell.

Daniels, P. W. and Moulaert, F. (eds) (1991) *The Changing Geography of Advanced Producer Services*, London: Belhaven.

Dear, M. J. (ed.) (2002) *From Chicago to LA: Making Sense of Urban Theory*, Thousand Oaks, CA: Sage.

Dematteis, G. (2000) 'Spatial images of European urbanisation', in A. Bagnasco and P. Le Gales (eds) *Cities in Contemporary Europe*, Cambridge: Cambridge University Press.

Derudder, B. and Taylor, P. J. (2003) 'The cliquishness of world cities', *GaWC Research Bulletin*, no. 113.

Derudder, B. and Witlox, T. (2002) 'Fuzzy classifications in large databases', *solstice*, 13(1): 32.

Derudder, B., Taylor, P. J., Witlox, F. and Catalano, G. (forthcoming) 'Hierarchical tendencies and regional patterns in the world city network: a global urban analysis of 234 cities', *Regional Studies*.

Dicken, P. (1998) *Global Shift*, 3rd edn, London: Paul Chapman.

Dollinger, P. (1970) *The German Hansa*, London: Macmillan.

Douglas, M. and Friedmann, J. (eds) (1998) *Cities for Citizens*, New York: Wiley.

Downs, A. (1991) 'Obstacles in the future of US cities', *Journal of the American Institute of Planners*, 57: 13–15.

Drbohlev, D. and Sykora, L. (1999) 'Gateway cities in the process of regional integration in Central and Eastern Europe: the case of Prague', in G. Buffl (ed.) *Migration, Free Trade and Regional Integration in Central and Eastern Europe*, Vienna: Staatsdrucherei AG.

Drennan, M. P. (1992) 'Gateway cities: the metropolitan sources of US producer service exports', *Urban Studies*, 29: 217–35.

Drennan, M. P. (1996) 'The dominance of international finance by London, New York and Tokyo', in P. W. Daniels and W. F. Lever (eds) *The Global Economy in Transition*, London: Longman.

Edvardsson, B., Edvinsson, L. and Nystrom, T. H. (1993) 'Internationalization in service companies', *Service Industries Journal*, 13: 80–97.

Esparaza, A. X. and Krmenec, A. J. (1994) 'Producer services trade in city systems: evidence from Chicago', *Urban Studies*, 31: 29–46.

Esparaza, A. X. and Krmenec, A. J. (1999) 'Entrepreneurship and extra-regional trade in producer services', *Growth and Change*, 30: 216–36.

Esparaza, A. X. and Krmenec, A. J. (2000) 'Large city interaction in the US urban system', *Urban Studies*, 37: 691–709.

Ettlinger, N. and Archer, J. C. (1987) 'City-size distributions and the world urban system in the twentieth century', *Environment and Planning A*, 19: 1161–74.

Falk, R. (2000) 'The quest for humane governance in an era of globalization', in D. Kalb, M. van der Land, R. Staring and N. Wilterdink (eds) *The Ends of Globalization*, Lanham, MD: Rowman & Littlefield.

Farnstein, S., Gordon, I. and Harloe, M. (1992) *Divided Cities: New York and London in the Contemporary World*, Oxford: Blackwell.

Feagin, J. R. and Smith, M. P. (1987) 'Cities and the new international division of labor', in M. P. Smith and J. R. Feagin (eds) *The Capitalist City*, Oxford: Blackwell.

Finnie, G. (1998) 'Wired cities', *Communications Week International*, 18 May: 19–22.

Forer, P. (1978) 'A place for plastic space?', *Progress in Human Geography*, 2: 230–67.

Foucher, M. (1987) 'Geographical approaches to the "Mediterranean Basin" of America', in P. Girot and E. Kofman (eds) *International Geopolitical Analysis: A Selection from Herodate*, London: Croom Helm.

Frank, A. G. (1969) *Latin America: Underdevelopment or Revolution*, New York: Monthly Review.

Frank, A. G. (1998) *Reorientate: Global Economy in the Asian Age*, Berkeley: University of California Press.

Friedmann, J. (1978) 'The spatial organization of power in the development of urban systems', in L. S. Bourne and J. W. Simmons (eds) *Systems of Cities*, New York: Oxford University Press.

Friedmann, J. (1986) 'The world city hypothesis', *Development and Change*, 17: 69–83.

Friedmann, J. (1995) 'Where we stand: a decade of world city research', in P. L. Knox and P. J. Taylor (eds) *World Cities in a World-System*, Cambridge: Cambridge University Press.

Friedmann, J. (2001) 'Intercity networks in a globalizing era', in A. J. Scott (ed.) *Global City-Regions*, New York: Oxford University Press.

Friedmann, J. and Wolff, G. (1982) 'World city formation: an agenda for research and action', *International Journal of Regional and Urban Research*, 3: 309–44.

Frobel, F., Heinrichs, J. and Kreye, O. (1979) *The New International Division of Labor*, New York: Cambridge University Press.

Fuller, S. S. (1989) 'The internationalisation of the Washington, D.C., area economy', in R. V. Knight and G. Gappert (eds) *Cities in a Global Society*, London: Sage.

Ganz, A. and Konga, L. F. (1989) 'Boston in the world economy', in R. V. Knight and G. Gappert (eds) *Cities in a Global Society*, London: Sage.

Geddes, P. (1924) 'A world league of cities', *Sociological Review*, 26: 166–7.

Giddens, A. (1998) *The Third Way*, Cambridge: Polity Press.

Giddens, A. (1999) *Runaway World*, Cambridge: Polity Press.

Godfrey, B. J. and Zhou, Y. (1999) 'Ranking world cities: multinational corporations and the global urban hierarchy', *Urban Geography*, 20: 268–81.

Gordon, P. and Richardson, H. W. (1998) 'World cities in North America: structural change and future challenges', in F.-C. Lo and Y.-M. Yeung (eds) *Globalization and the World of Large Cities*, Tokyo: United Nations University Press.

Gottmann, J. (1961) *Megalopolis*, New York: Twentieth Century Fund.

Gottmann, J. (1984) *Orbits: The Ancient Mediterranean Tradition of Urban Networks*, Oxford: Leopard's Head Press.

Gottmann, J. (1987) *Megalopolis Revisited: 25 Years Later*, College Park, MD: Institute of Urban Studies, University of Maryland.

Gottmann, J. (1989) 'What are cities becoming centres of? Sorting out the possibilities', in R. V. Knight and G. Gappert (eds) *Cities in a Global Society*, Newbury Park, CA: Sage.

Graham, S. (1999) 'Global grids of glass', *Urban Studies*, 36: 929–49.

Graham, S. and Marvin, S. (1996) *Telecommunications and the City*, London: Routledge.

Hall, P. (1966) *The World Cities*, London: Heinemann.

Hall, P. and Hay, D. (1978) *Growth Centres in the European Urban System*, London: Heinemann.

Hansen, N. (2000) 'Miami: multicultural gateway of the Americas', in A. E. Andersson and D. E. Andersson (eds) *Gateways to the Global Economy*, Cheltenham, UK: Edward Elgar.

Harris, C. D. and Ullman, E. L. (1945) 'The nature of cities', *Annals of the American Academy of Political and Social Science*, 242: 7–17.

Harris, N. (1997) 'Cities in a global economy: structural change and policy reactions', *Urban Studies*, 34: 1693–1703.

Harvey, D. (1990) *The Condition of Postmodernity*, Oxford: Blackwell.

Harvey, T. (1996) 'Portland, Oregon: regional city in a global economy', *Urban Geography*, 17: 95–114.

Held, D., McGrew, A., Goldblatt, D. and Perraton, J. (1999) *Global Transformations*, Cambridge, UK: Polity Press.

Hicks, D. A. and Nivin, S. R. (1996) 'Minneapolis-St Paul in the global economy', *Urban Geography*, 17: 23–42.

Hill, R. C. and Fujita, K. (1995) 'Osaka's Tokyo problem', *International Journal of Urban and Regional Research*, 19: 181–91.

Hinsley, F. H. (1982) 'The rise and fall of the modern international system', *Review of International Studies*, 8: 1–8.

Hopkins, T. K. and Wallerstein, I. (1996) *The Age of Transition*, London: Pluto.

Hymer, S. (1972) 'The multinational corporation and the law of uneven development', in J. Bhagwati (ed.) *Economics and World Order from the 1970s to the 1990s*, London: Collier-Macmillan.

Isard, W. (1956) *Location and Space-Economy*, New York: Wiley.

Islin, E. F. (2000) 'Introduction: democracy, citizenship and the city', in E. F. Islin (ed.) *Democracy, Citizenship and the Global City*, London: Routledge.

Israel, J. I. (1989) *Dutch Primacy in World Trade, 1585–1740*, Oxford: Clarendon Press.

Israel, J. I. (1995) *The Dutch Republic*, Oxford: Clarendon Press.

Jacobs, J. (1984) *Cities and the Wealth of Nations*, New York: Vintage.

Johnson, J. H. (1967) *Urban Geography*, Oxford: Pergamon.

Johnston, R. J. (1982) *The American Urban System: A Geographical Perspective*, London: Longman.

Jones, A. (1998) 'Re-theorising the core: a "globalised" business elite in Santiago, Chile', *Political Geography*, 17: 295–318.

Keeling, D. J. (1995) 'Transport and the world city paradigm', in P. L. Knox and P. J. Taylor (eds) *World Cities in a World-System*, Cambridge: Cambridge University Press.

King, A. D. (1990) *Global Cities*, London: Routledge.

Knight, R. V. (1989a) 'The emergent global society', in R. V. Knight and G. Gappert (eds) *Cities in a Global Society*, Newbury Park, CA: Sage.

Knight, R. V. (1989b) 'City building in a global society', in R. V. Knight and G. Gappert (eds) *Cities in a Global Society*, Newbury Park, CA: Sage.

Knight, R. V. and Gappert, G. (eds) (1989) *Cities in a Global Society*, Newbury Park, CA: Sage.

Knoke, D. and Kuklinski, J. H. (1982) *Network Analysis*, Beverly Hills, CA: Sage.

Knox, P. L. (1995) 'World cities in a world-system', in P. L. Knox and P. J. Taylor (eds) *World Cities in a World-System*, Cambridge: Cambridge University Press.

Knox, P. L. (1996) 'Globalization and urban change', *Urban Geography*, 17: 115–17.

Knox, P. L. and Agnew, J. (1989) *The Geography of the World Economy*, London: Arnold.

Knox, P. L. and Taylor, P. J. (eds) (1995) *World Cities in a World-System*, Cambridge: Cambridge University Press.

Korff, R. (1987) 'The world city hypothesis: a critique', *Development and Change*, 17: 483–95.

Kossmann, E. H. and Mellick, A. F. (eds) (1974) *Texts Concerning the Revolt of the Netherlands*, Cambridge: Cambridge University Press.

Kratke, S. (2002) *Medienstadt*, Opladen: Leske & Budrich.

Kratke, S. and Taylor, P. J. (forthcoming) 'A world geography of global media cities', *European Planning Studies*.

Kresl, P. K. (1995) 'The determinants of urban competitiveness: a survey', in P. K. Kresl and G. Gappert (eds) *North American Cities and the Global Economy*, Thousand Oaks, CA: Sage.

Kresl, P. K. and Gappert, G. (eds) (1995) *North American Cities and the Global Economy*, Thousand Oaks, CA: Sage.

Kresl, P. K. and Singh, B. (1999) 'Competitiveness and the urban economy in twenty-four large US metropolitan areas', *Urban Studies*, 36: 1017–27.

Krugman, P. R. (1994) 'Competitiveness: a dangerous obsession', *Foreign Affairs*, 73(2): 28–44.

Kruskal, J. B. and Wish, M. (1978) *Multidimensional Scaling*, Beverly .Hills, CA: Sage.

Kunzmann, K. R. (1998) 'World city regions in Europe: structural change and future challenges', in F.-C. Lo and Y.-M. Yeung (eds) *Globalization and the World of Large Cities*, Tokyo: United Nations University Press.

Lakshmanan, T. R., Andersson, D. E., Chatterjee, L. and Sasaki, K. (2000) 'Three global cities: New York, London and Tokyo', in A. E. Andersson and D. E. Andersson (eds) *Gateways to the Global Economy*, Cheltenham, UK: Edward Elgar.

Lee, R. and Pelizzon, S. (1991) 'Hegemonic cities in the modern world-system', in R. Kasaba (ed.) *Cities in the World-System*, New York: Greenwood.

Leslie, D. A. (1995) 'Global scan: the globalization of advertising agencies, concepts, and campaigns', *Economic Geography*, 71: 402–26.

Lever, W. F. (1999) 'Competitive cities in Europe', *Urban Studies*, 36: 1029–44.

Lever, W. F. and Turok, I. (1999) 'Competitive cities: introduction to the review', *Urban Studies*, 36: 791–3.

Lewis, M. W. and Wigen, K. E. (1997) *The Myth of Continents*, Berkeley: University of California Press.

Li, J. and Guisinger, S. (1992) 'The globalization of service multinationals in the "Triad" regions: Japan, Western Europe and North America', *Journal of International Business Studies*, 23: 145–61.

Lloyd, T. H. (1991) *England and the German Hanse, 1157–1611*, Cambridge: Cambridge University Press.

Lo, F.-C. and Yeung, Y.-M. (eds) (1998) *Globalization and the World of Large Cities*, Tokyo: United Nations University Press.

Logan, J. R. (2000) 'Still a global city: the racial and ethnic segmentation of New York', in P. Marcuse and R. van Kempen (eds) *Globalizing Cities*, Oxford: Blackwell.

London Planning Advisory Council (1991) *London: World City Moving into the Twenty First Century*, London: HMSO.

Lukermann, F. (1966) 'Empirical expressions of nodality and hierarchy in a circulation manifold', *East Lakes Geographer*, 2: 17–44.

Lyons, D. and Salmon, S. (1995) 'World cities, multinational corporations, and urban hierarchy: the case of the United States', in P. L. Knox and P. J. Taylor (eds) *World Cities in a World-System*, Cambridge: Cambridge University Press.

McCormick, B. H., DeFanti, T. A. and Brown, M. D. (1987) 'Visualization in scientific computing', *Computer Graphics*, 21: 23–45.

Mackay, J. R. (1958) 'The interactance hypothesis and boundaries in Canada: a preliminary study', *Canadian Geographer*, 11: 1–8.

Malecki, E. J. (1980) 'Science and technology in the American urban system', in S. D. Brunn and J. O. Wheeler (eds) *The American Metropolitan System*, New York: Wiley.

Malecki, E. J. (2002) 'Hard and soft networks for urban competitiveness', *Urban Studies*, 39: 929–46.

Marcuse, P. and van Kempen, R. (eds) (2000) *Globalizing Cities*, Oxford: Blackwell.

Markusen, A. and Gwiasda, V. (1994) 'Multipolarity and the layering of functions in world cities: New York City's struggle to stay on top', *International Journal of Urban and Regional Research*, 18: 167–93.

Martin, H.-P. and Schumann, H. (1997) *The Global Trap*, London: Zed.

Massey, D. (1999) 'Cities in the world', in D. Massey, J. Allen and S. Pile (eds) *City Worlds*, London: Routledge.

Massey, D., Allen, J. and Pile, S. (eds) (1999a) *City Worlds*, London: Routledge.

Massey, D., Allen, J. and Pile, S. (1999b) 'Introduction', in D. Massey, J. Allen and S. Pile (eds) *City Worlds*, London: Routledge.

Mayerhofer, P. and Wolfmayr-Schnitzer, Y. (1997) 'Gateway cities in the process of regional integration in Central and Eastern Europe', in G. Buffl (ed.) *Migration, Free Trade and Regional Integration in Central and Eastern Europe*, Vienna: Staatsdrucherei AG.

Michelson, R. L. and Wheeler, J. O. (1994) 'The flow of information in a global economy: the role of the American urban system in 1990', *Annals of the Association of American Geographers*, 84: 87–107.

Misra, J. and Boswell, T. (1997) 'Dutch hegemony during the age of mercantilism', *Acta Politica*, 32: 174–209.

Modelski, G and Thompson, W. (1988) *Sea Power in Global Politics, 1494–1993*, Seattle: University of Washington Press.

Mollenkopf, J. and Castells, M. (eds) (1991) *Dual City: Restructuring New York*, New York: Russell Sage Foundation.

Morata, F. (1997) 'The Euro-region and the C-6 network: the new politics of sub-national co-operation in the western Mediterranean area', in M. Keating and J. Loughlin (eds) *The Political Economy of Regionalism*, London: Frank Cass.

Nijman, J. (1996) 'Breaking the rules: Miami in the urban hierarchy', *Urban Geography*, 17: 5–22.

Nijman, J. (1997) 'Globalization to a Latin beat: the Miami growth machine', *Annals of the American Academy of Political and Social Sciences*, 551: 163–76.

O'Brien, P., Keene, D., t'Hart, M. and van der Wee, H. (eds) (2001) *Urban Achievements in Early Modern Europe*, Cambridge: Cambridge University Press.

O'Farrell, P. N., Wood, P. A. and Zheng, J. (1996) 'Internationalization of business services: an inter-regional analysis, *Regional Studies*, 30: 101–18.

Oliver, R. and Fage, J. D. (1988) *A Short History of Africa*, London: Penguin.

Orford, S., Dorling, D. and Harris, R. (1998) *Review of Visualization in the Social Sciences: A State of the Art Survey and Report*, Bristol: School of Geographical Sciences, University of Bristol.

Pedersen, P. O. (1978) 'Innovation diffusion within and between national urban systems', in L. S. Bourne and J. W. Simmons (eds) *Systems of Cities*, New York: Oxford University Press.

Petrella, R. (1995) 'A global agora vs. gated city-regions', *New Perspectives*, Winter: 21–2.

Phillips, P. D. and Brunn, S. D. (1980) 'New dynamics of growth in the American metropolitan system', in S. D. Brunn and J. O. Wheeler (eds) *The American Metropolitan System: Present and Future*, New York: Wiley.

Pocock, J. G. A. (1992) 'The Dutch republican tradition' in M. C. Jacob and W. W. Mijnhardt (eds) *The Dutch Republic in the Eighteenth Century*, Ithaca, NY: Cornell University Press.

Poon, J. P. H. (2000) 'Reconfiguring regional hierarchy through regional offices in Singapore', in A. E. Andersson and D. E. Andersson (eds) *Gateways to the Global Economy*, Cheltenham, UK: Edward Elgar.

Powell, W. W. (1990) 'Neither market nor hierarchy: network forms of organization', *Research in Organizational Behavior*, 12: 295–336.

Pred, A. (1977) *City-Systems in Advanced Economies*, London: Hutchinson.

Pred, A. (1978) 'On the spatial structure of organizations and the complexity of metropolitan interdependence', in L. S. Bourne and J. W. Simmons (eds) *Systems of Cities*, New York: Oxford University Press.

Princen, T. and Finger, M. (1994) 'Introduction', in T. Princen and M. Finger (eds) *Environmental NGOs in World Politics*, London: Routledge.

Rakodi, C. (1998) 'Globalization trends and sub-Saharan African cities', in F.-C. Lo and Y.-M. Yeung (eds) *Globalization and the World of Large Cities*, Tokyo: United Nations University Press.

Reed, H. C. (1981) *The Pre-eminence of International Financial Centers*, New York: Praeger.

Rimmer, P. J. (1991) 'The global intelligence corps and world cities: engineering consultancies on the move', in P. W. Daniels (ed.) *Services and Metropolitan Development: International Perspectives*, London: Routledge.

Rimmer, P. J. (1998) 'Transport and telecommunications among world cities', in F.-C. Lo and Y.-M. Yeung (eds) *Globalization and the World of Large Cities*, Tokyo: United Nations University Press.

Ringrose, D. (1989) 'Towns, transport and crown: geography and the decline of Spain', in E. D. Genovese and L. Hochberg (eds) *Geographic Perspectives in History*, Oxford: Blackwell.

Roberts, J. (1999) 'The internationalisation of business service firms', *Service Industries Journal*, 19: 68–88.

Rondinelli, D. A., Johnson, J. H. Jr and Kasarda, J. D. (1998) 'The changing forces of urban economic development: globalization and city competitiveness in the 21st century', *Cityscape*, 3: 71–105.

Rostow, W. W. (1960) *Stages of Economic Growth*, Cambridge: Cambridge University Press.

Rowen, H. H. (1978) *The Princes of Orange*, Cambridge: Cambridge University Press.

Ruggie, J. (1993) 'Territoriality and beyond', *International Organization*, 47: 139–74.

Rummel, R. J. (1970) *Applied Factor Analysis*, Evanston, IL: Northwestern University Press.

Sassen, S. (1991) *The Global City*, Princeton, NJ: Princeton University Press.

Sassen, S. (1993) 'Miami: a new global city?', *Contemporary Sociology*, 22: 471–7.

Sassen, S. (1994) *Cities in a World Economy*, Thousand Oaks, CA: Pine Forge.

Sassen, S. (1999) 'A new emergent hegemonic structure? *Political Power and Social Theory*, 13: 277–89.

Sassen, S. (2000) 'New frontiers facing urban sociology at the millennium', *British Journal of Sociology*, 51: 143–59.

Sassen, S. (2001) *The Global City*, 2nd edn, Princeton, NJ: Princeton University Press.

Sassen, S. (ed.) (2002) *Global Networks, Linked Cities*, London: Routledge.

Scanlon, R. (1989) 'New York City as global capital in the 1980s', in R. V. Knight and G. Gappert (eds) *Cities in a Global Society*, London: Sage.

Schama, S. (1987) *The Embarrassment of Riches*, London: Collins.

Scott, A. J. and Soja, E. W. (eds) (1998) *The City: Los Angeles and Urban Theory at the End of the Twentieth Century*, Berkeley: University of California Press.

Scott, J. (1991) *Social Network Analysis*, London: Sage.

Shefter, M. (1993) *Capital of the American Century: The National and International Influence of New York City*, New York: Russell Sage Foundation.

Shin, K.-H., and Timberlake, M. (2000) 'World cities in Asia: cliques, centrality and connectedness', *Urban Studies*, 37: 2257–85.

Short, J. R. and Kim, Y.-H. (1999) *Globalization and the City*, London: Longman.

Short, J. R., Kim, Y., Kuss, M. and Wells, H. (1996) 'The dirty little secret of world city research', *International Journal of Regional and Urban Research*, 20: 697–717.

Short, J. R., Breitbach, C., Buckman, S. and Essex, J. (2000) 'From world cities to gateway cities', *City*, 4: 317–40.

Silver, B. J. (1995) 'World-scale patterns of capital–labor conflict', *Review* (Fernand Braudel Center), 18: 155–92.

Simmons, J. W. (1978) 'The organization of the urban system', in L. S. Bourne and J. W. Simmons (eds) *Systems of Cities*, New York: Oxford University Press.

Sklair, L. (1998) 'The transnational capitalist class and global capitalism', *Political Power and Social Theory*, 12: 3–43.

Smith, A. D. (1982) 'Ethnic identity and world order', *Millennium*, 12: 149–61.

Smith, D. and Timberlake, M. (2002) 'Hierarchies of dominance among world cities: a network approach', in S. Sassen (ed.) *Global Networks, Linked Cities*, London: Routledge.

Smith, M. P. (2001) *Transnational Urbanism*, Oxford: Blackwell.

Smith, W. D. (1984) 'The function of commercial centres in the modernisation of European capitalism: Amsterdam as an information exchange in the seventeenth century', *Journal of Economic History*, 44: 985–1005.

Soja, E. W. (2000) *Postmetropolis*, Oxford: Blackwell.

Soja, E. W., Morales, R. and Wolff, G. (1983) 'Urban restructuring: an analysis of urban and spatial change in Los Angeles', *Economic Geography*, 59: 195–230.

Song, S. and Zhang, K. H. (2002) 'Urbanization and city size distribution in China', *Urban Studies*, 39: 2317–27.

Spar, D. L. (1997) 'Lawyers abroad: the internationalisation of legal practice', *California Management Review*, 39: 8–28.

Spufford, P. (2002) *Power and Profit: The Merchant in Medieval Europe*, London: Thames & Hudson.

Stanback, T. M. and Noyelle, T. J. (1982) *Cities in Transition*, Totowa, NJ: Allanheld, Osmun.

Stephens, J. D. and Holly, B. P. (1980) 'The changing patterns of industrial corporate control in the metropolitan United States', in S. D. Brunn and J. O. Wheeler (eds) *The American Metropolitan System: Present and Future*, New York: Wiley.

Stough, R. R. (2000) 'The Greater Washington region: a global gateway region', in A. E. Andersson and D. E. Andersson (eds) *Gateways to the Global Economy*, Cheltenham, UK: Edward Elgar.

Suarez-Villa, L. (2000) 'Southern California as a global gateway region: polycentricity and network segmentation as competitive advantage', in A. E. Andersson and D. E. Andersson (eds) *Gateways to the Global Economy*, Cheltenham, UK: Edward Elgar.

Taaffe, E., Morrill, R. L. and Gould, P. R. (1963) 'Transport expansion in underdeveloped countries: a comparative analysis', *Geographical Review*, 53: 503–29.

Taylor, P. J. (1982) 'A materialist framework for political geography', *Transactions of the Institute of British Geographers*, NS 7: 15–34.

Taylor, P. J. (1991) 'The English and their Englishness', *Scottish Geographical Magazine*, 107: 146–61.

Taylor, P. J. (1994) 'Ten years that shook the world? The United Provinces as the first hegemonic state', *Sociological Perspectives*, 37: 25–46.

Taylor, P. J. (1995) 'World cities and territorial states: the rise and fall of their mutuality', in P. L. Knox and P. J. Taylor (eds) *World Cities in a World-System*, Cambridge: Cambridge University Press.

Taylor, P. J. (1996) *The Way the Modern World Works: World Hegemony to World Impasse*, Chichester, UK: Wiley.

Taylor, P. J. (1997) Hierarchical tendencies amongst world cities: a global research proposal', *Cities*, 14: 323–32.

Taylor, P. J. (1999) *Modernities: A Geohistorical Interpretation*, Cambridge: Polity Press.

Taylor, P. J. (2000a) 'World cities and territorial states under conditions of contemporary globalization', *Political Geography*, 19: 5–32.

Taylor, P. J. (2000b) 'Embedded statism and the social sciences 2: Geographies (and metageographies) in globalization', *Environment and Planning A*, 32: 1105–14.

Taylor, P. J. (2000c) 'Izations of the world: Americanization, modernization and globalization', in C. Hay and D. Marsh (eds) *Demystifying Globalization*, London: Macmillan.

Taylor, P. J. (2001a) 'Specification of the world city network', *Geographical Analysis*, 33: 181–94.

Taylor, P. J. (2001b) 'Being economical with the geography', *Environment and Planning A*, 33: 949–54.

Taylor, P. J. (2003) 'Metageographical moments: a geohistorical interpretation of embedded statism and globalization', in M. A. Tetreault, R. A. Denmark, K. P. Thomas and K. Burch (eds) *Rethinking Political Global Economy: Emerging Issues, Unfolding Odysseys*, London: Routledge.

Taylor, P. J. and Hoyler, M. (2000) 'The spatial order of European cities under conditions of contemporary globalization', *Tijdschrift voor Economische en Sociale Geografie*, 91: 176–89.

Taylor, P. J. and Walker, D. R. F. (forthcoming) 'Hinterlands revisited', *Geography*.

Taylor, P. J., Catalano, G. and Walker, D. R. F. (2002a) 'Measurement of the world city network', *Urban Studies*, 39: 2367–76.

Taylor, P. J., Catalano, G. and Walker, D. R. F. (2002b) 'Exploratory analysis of the world city network', *Urban Studies*, 39: 2377–94.

Taylor, P. J., Catalano, G. and Walker, D. R. F. (forthcoming) 'Multiple globalisations: regional, hierarchical and sectoral articulations of global business services through world cities', *Services Industries Journal*.

Taylor, P. J., Walker, D. R. F., Catalano, G. and Hoyler, M. (2002c) 'Diversity and power in the world city network', *Cities*, 19: 231–41.

t'Hart, M. (1989) 'Cities and statemaking in the Dutch Republic, 1580–1680', *Theory and Society*, 18: 663–87.

t'Hart, M. (1993) *The Making of a Bourgeois State*, Manchester: Manchester University Press.

Thrift, N. (1987) 'The fixers? The urban geography of international commercial capital', in J. Henderson and M. Castells (eds) *Restructuring and Territorial Development*, Beverly Hills, CA: Sage.

Thrift, N. (1999) 'Cities and economic change: global governance?', in J. Allen, D. Massey and M. Pryke (eds) *Unsettling Cities*, London: Routledge.

Townsend, A. M. (2001) 'Network cities and the global structure of the internet', *American Behavioral Scientist*, 44: 1697–1716.

Vandermerwe, S. and Chadwick, M. (1989) 'The internationalisation of services', *The Service Industries Journal*, 9: 79–93.

van der Pijl, K. (2002) 'Holding the middle ground in the transnationalisation process', in J. Anderson (ed.) *Transnational Democracy*, London: Routledge.

Wallerstein, I. (1979) *The Capitalist World-Economy*, Cambridge: Cambridge University Press.

Wallerstein, I. (1980) *The Modern World-System II*, New York: Academic Press.

Wallerstein, I. (1984) *Politics of the World-Economy*, Cambridge: Cambridge University Press.

Walton, R. (1996) 'Another round of globalization in San Francisco', *Urban Geography*, 17: 60–94.

Warf, B. and Erickson, R. (1996) 'Introduction: globalization and the US city system', *Urban Geography*, 17: 1–4.

Wilson, C. (1957) *Profit and Power: A Study of England and the Dutch Wars*, London: Longman Green.

Wilson, C. (1958) *Mercantilism*, London: Historical Association.

Wood, P. A. (1987) 'Producer services and economic change', in K. Chapman and G. Humphry (eds) *Technological Change and Industrial Policy*, Oxford: Blackwell.

Wu, W. (1996) 'Economic competition and resource mobilization', in M. A. Cohen, B. A. Ruble, J. S. Tulchin and A. M. Garland (eds) *Preparing for the Urban Future*, Washington, DC: Woodrow Wilson Center Press.

Young, F. W. (1987) *Multidimensional Scaling: History, Theory and Applications*, Hillsdale, NJ: Lawrence Erlbaum Associates.

Zerubavel, E. (1992) *Terra Cognita: The Mental Discovery of America*, New Brunswick, NJ: Rutgers University Press.

Zook, M. A. (2001) 'Old hierarchies or new networks of centrality?', *American Behavioral Scientist*, 44: 1679–96.

Index

Note: Where country names are indexed see also entries under their individual cities.

eBooks

eBooks - at www.eBookstore.tandf.co.uk

A library at your fingertips!

eBooks are electronic versions of print books. You can store them onto your PC/laptop or browse them online.

They have advantages for anyone needing rapid access to a wide variety of published, copyright information.

eBooks can help your research by enabling you to bookmark chapters, annotate and use instant searches to find specific words or phrases. Several eBook files would fit on even a small laptop or PDA.

NEW: Save money by eSubscribing: cheap, online acess to any eBook for as long as you need it.

Annual subscription packages

We now offer special low cost bulk subscriptions to packages of eBooks in certain subject areas. These are available to libraries or to individuals.

For more information please contact webmaster.ebooks@tandf.co.uk

We're continually developing the eBook concept, so keep up to date by visiting the website.

www.eBookstore.tandf.co.uk